Turbine Science and Technology

Turbomachinery Clearance Control

Raymond E. Chupp*

General Electric Global Research

Niskayuna, New York 12302

Robert C. Hendricks†

National Aeronautics and Space Administration

Glenn Research Center

Cleveland, Ohio 44135

Scott B. Lattime‡

The Timken Company

North Canton, Ohio 44720

Bruce M. Steinetz§

National Aeronautics and Space Administration

Glenn Research Center

Cleveland, Ohio 44135

Mahmut F. Aksit¶

Sabanci University

Istanbul, Turkey

*Mechanical Engineer, Energy and Propulsion Technologies, One Research Circle, AIAA Member.

†Senior Technologist Research and Development Directorate, 21000 Brookpark Road, AIAA Member.

‡Tribolologic Specialist, 1835 Dueber Ave. SW, P.O Box 6932.

§Senior Technologist, Materials and Structures Division, 21000 Brookpark Road, Associate Fellow.

¶Faculty of Engineering and Natural Sciences, Orhanli 34956 Tuzla.

Acronym List

ACC	active clearance control
APS	air plasma spray
BOM	bill of material
EB-PVD	electron beam plasma vapor deposition
FEA	finite element analysis
FOD	foreign object damage
HCF	high-cycle fatigue
FSN	first-stage nozzle
HFBS	hybrid floating brush seal
HPC	high-pressure compressor
HPP	high-pressure packing
HPT	high-pressure turbine

IPC	intermediate pressure compressor
IPT	intermediate pressure turbine
LCF	low-cycle fatigue
LDV	laser Doppler velocimetry
LPC	low-pressure compressor
LPT	low-pressure turbine
LSAC	low-speed axial compressor
MTBF	mean time between failures
OSD	oxide dispersion strengthened
PSZ	partially-stabilized zirconia
T1	first turbine stage
TD	transition duct
TIR	total indicator reading
TO	takeoff
YSZ	yttria-stabilized zirconia

I. Introduction

Controlling interface clearances is the most cost effective method of enhancing turbomachinery performance. Seals control turbomachinery leakages, coolant flows and contribute to overall system rotordynamic stability. In many instances, sealing interfaces and coatings are sacrificial, like lubricants, giving up their integrity for the benefit of the component. They are subjected to abrasion, erosion, oxidation, incursive rubs, foreign object damage (FOD) and deposits as well as extremes in thermal, mechanical, aerodynamic and impact loadings.

Tribological pairing of materials control how well and how long these interfaces will be effective in controlling flow.

A variety of seal types and materials are required to satisfy turbomachinery sealing demands. These seals must be properly designed to maintain the interface clearances. In some cases, this will mean machining adjacent surfaces, yet in many other applications, coatings are employed for optimum performance. Many seals are coating composites fabricated on superstructures or substrates that are coated with sacrificial materials which can be refurbished either in situ or by removal, stripping, recoating and replacing until substrate life is exceeded.

For blade and knife tip sealing an important class of materials known as abradables permit blade or knife rubbing without significant damage or wear to the rotating element while maintaining an effective sealing interface. Most such tip interfaces are passive, yet some, as for the high-pressure turbine (HPT) case or shroud, are actively controlled.

This work presents an overview of turbomachinery sealing. Areas covered include: characteristics of gas and steam turbine sealing applications and environments, benefits of sealing, types of standard static and dynamics seals, advanced seal designs, as well as life and limitations issues.

II. Sealing in Gas and Steam Turbines

A. Clearance Control Characteristics

Turbomachines range in size from centimeters (size of a penny) to ones you can almost walk through. The problem is how to control the large changes in geometry between adjacent rotor/stator components from cold-build to operation. The challenge is to provide geometric control while maintaining efficiency, integrity and long service life (e.g., estimated time to failure or maintenance, and low cost[1]). Figure 1 shows the relative clearance between the rotor tip and case for a HPT during takeoff, climb, and cruise conditions.[2] The figure shows the dramatic effect of clearance control via applied cooling to the casing. A critical clearance requirement occurs at "cut-back" (about 1000 s into climb-out) when takeoff thrust is reduced. Using thermal active clearance control (ACC), the running clearance is drastically reduced, producing significant cost savings in fuel reduction and increased service life. However, designers must note that changing parameters in critical seals can change the dynamics of the entire engine.[3] These effects are not always positive.

B. Sealing Benefits

Performance issues are closely tied to engine clearances. Ludwig[4] determined that improvements in fluid film sealing resulting from a proposed research program could lead to an annual energy saving, on a national basis, equivalent to about 37 million barrels (1.554 billion = 1554 million U.S. gallons) of oil or 0.3 percent of the total U.S. energy consumption (1977 statistics). In terms engine bleed, Moore[5] cited that a 1-percent reduction in engine bleed gives a 0.4-percent reduction in specific fuel consumption (SFC), which translates into nearly 0.033 (1977 statistics) to 0.055 (2004 statistics) billion gallons of U.S. airlines fuel savings and nearly 0.28 billion gallons world wide (2004 statistics), annually. In terms of clearance changes, Lattime and Steinetz[6] cite a 0.0254 mm (0.001 in.) change in HPT tip clearance, decreases SFC by 0.1 percent and EGT (exhaust gas temperature) by 1 °C, producing an annual savings of 0.02

billion gallons for U.S. airlines. In terms of advanced sealing, Munson et al.[7] estimate savings of over 0.5 billion gallons of fuel. Chupp et al.[8] estimated that refurbishing compressor seals would yield impressive improvements across the fleet ranging from 0.2 to 0.6 percent reduction in heat-rate and 0.3 to 1 percent increase in power output. For these large, land-based gas turbines, the percentages represent huge fuel savings and monetary returns with the greatest returns cited for aging power systems.

C. The Sealing Environment

1. Seal Types and Locations

Key aero-engine sealing and thermal restraint locations cited by Bill[9] are shown in figure 2. These include the fan and compressor shroud seals (rub strips), compressor interstage and discharge seals (labyrinth), combustor static seals, balance piston sealing, turbine shroud and rim-cavity sealing. Industrial engines have similar sealing requirements. Key sealing locations for the compressor and turbine in an industrial engine are cited by Aksit[10] and Camatti et al.[11,12] and are shown in figures 3 and 4, with an overview of sealing in large industrial gas turbines by Hurter,[13] figure 3(c).

Figure 3 shows high-pressure compressor (HPC) and HPT tip seal (abradable) and interstage seal (brush seal) locations, while figure 4 shows impeller shroud (labyrinth) and interstage seal (honeycomb) locations for the compressor. Compressor interstage platform seals are of the shrouded type (figs. 5 and 6). These seals are used to minimize backflow, stage pressure losses and re-ingested passage flow. Turbine stators, also of the shrouded type, prevent hot gas ingestion into the cavities that house the rotating disks and control blade and disk coolant flows.

Designers need to carefully consider the differences in thermal and structural characteristics, pressure gradient differences, and blade rub interfaces.

Characteristically the industrial gas turbine can be thought of as a heavy-duty derivative of an aero engine. Still industrial and aero-turbomachines have many differences. The most notable are the fan, spools and combustor. Aero engines derive a large portion of their thrust through the bypass fan and usually have inline combustors, high and low pressure spools, drum rotors and high exhaust velocities, all subject to flight constraints. Large industrial engines (fig. 7) have plenum inlets, can-combustors, single spools, through-bolted-stacked disc rotors and exhaust systems constrained by 640 °C (1180 °F) combined cycle (steam-reheat-turbine) requirements. In both types of engines, core requirements are similar, yet materials restraints differ.

2. *Materials and Environmental Conditions*

Over the years, advances in new base materials, notably Ni-based single crystal alloys, and coatings have allowed increased operating temperatures of turbine engine components. Complementary to the thermal and pressure profiles, materials used range from steel to superalloys coated with metallics and ceramics. Variations in engine pressure and temperature of the Rolls-Royce Trent gas turbine[*] are illustrated in figure 8 which also has an intermediate pressure turbine (IPT). The lower temperature blades in the fan and low-pressure compressor (LPC) sections are made of titanium, or composite materials, with corrosion resistant coatings

---

[*](Data available from Sourmail, T., "Coatings for Turbine Blades," University of Cambridge, at http://www.msm.cam.ac.uk/phase-trans/2003/Superalloys/coatings/ [cited 18 May 2005].),

due to their high strength and low density. The elevated temperatures of the HPC, HPT, and low-pressure turbine (LPT) require the use of Nickel-based superalloys. In the HPT of aero-engines, for example, the first-stage turbine blades can see gas path temperatures around 1400 °C (2550 °F). To withstand these punishing temperatures for the 20,000-hr and more service lives, aero-engine designers have turned to single crystal blades, with thermal barrier coatings generally using yttria-stabilized zirconia (YSZ).[2] In this configuration, blade metal temperatures reach 982 °C (1800 °F) and ceramic surface temperatures reach 1100 °C (2010 °F).

To improve blade tip sealing effectiveness, squealer tips (approximately 0.8mm (0.03-in.) high) are integrated into the blade. Depending on engine design, adjacent shroud seals are made of either directionally solidified cast superalloy materials coated with sprayed abradable coatings (YSZ based)[14] or single crystal shroud segments capable of the required operating temperatures.

III. Static Sealing

Sealing at static or slow-relative-motion interface locations in turbomachinery includes the sealing at interfaces or junctions between the stationary components (combustors, nozzles, shrouds, and structure, etc.) throughout the internal cooling flow path to minimize or control parasitic leakage flows between turbine components (see Figure 3(b)). Typically, adjacent members have to sustain relative vibratory motion with minimal wear or loss of sealing over the design life of the seal. In addition, these types of seals must be compliant to accommodate thermal growth and misalignment. For example, in a cooling application, static (and some structural) seals provide a careful balance of leakage, cooling, and to some extent, dynamics. If the seals are too tight, improper damping occurs, particularly with low differential pressure loads. Coolant leakage flows then cannot properly purge cavities or offer sufficient cooling to protect themselves against ingested hot gas streaks. Rapid detoriation, oxidizing, and burning can occur. If coolant-seal leakage is excessive, coolant air leaking into the powerstream introduces parasitic air loss and low energy fluid that increases the average passage blockage factor (see also Appendix A). Effective sealing at these static interface locations not only increases turbine efficiency and power output, but also improves the main-gas-path temperature profile. Various compliant-interface seals have been developed to address these issues and are discussed in the following sections.

A. Metallic Seals

For smaller gap movements, more conventional seals are used. These seals are metallic for higher temperature and pressure environments where rubber and polymer seals are not suitable. The wide range of applications in turbomachinery drives the need for multiple configurations, such as the O, C, and E-type cross section (see fig. 9). The type of seal that is best suited for a particular application depends on operating variables such as temperature, pressure, required leakage rate, flange separation, fatigue life, and the load available to seat the seal. Figure 10 shows an example of where some metal seals are used in an industrial gas turbine. There are many smaller "feather" seals (thin sheet metal) used throughout; all interfaces require sealing of some nature.

In higher temperature environments, a large amount of thermal growth in surrounding structures is typical. This makes it necessary for the metal seal to maintain contact with the sealing surfaces while the structure moves. A seal's ability to follow the moving structure is due to its spring back and system pressure to seat the seal. In general, E-type seals (alternatively called W-seals) provide the largest amount of spring back. For this reason, the majority of metal seals found in steam, gas, and jet engines are of the E-type configuration.

The ability of a seal to maintain a low leakage rate is mostly caused by the force the seal exerts on the mating flange, also called the "seating load." Typically speaking, the leakage rate of a seal will decrease as the seal seating load increases. C-type seals have higher seating loads than E-type seals. To further increase a C-type seal load and spring back, a spring can be inserted around the circumference on the inside of the cross section. The high load of a C seal can be used to enhance sealing performance by the addition of plating such as silver, nickel, gold and copper. The simple geometry of a C-type seal limits further design possibilities.

The relative complexity and adaptability of the E-type seal cross sections allows for increased design variations with somewhat increased leakage rates compared to C-seals. The number of convolutions, material thickness, convolution depth and free height all play a major roll in seal performance. Despite the large thermal growth common in turbine engines, a properly designed E-type seal can have millions-of-cycles fatigue life. A majority of the E-type seals used in turbine engines are located between engine case segments, such as the horizontal joint in the combustion section on steam and gas turbines. E-type seals can also be found in the cartridge assemblies of a turbine fuel nozzle. The seal can be cut axially in one or more circumferential locations to accommodate radial growth difference or assembly requirements. Small "caps" can be placed on the seal to span the circumferential gaps to control leakage.

Currently metallic seals for higher temperature applications are made from Inconel 718 and Waspaloy nickel-based superalloys with a temperature limit of about 730 °C (1350 °F).[15] Above this temperature, such seals under compressive and tensile stresses relax due to creep with an attendant loss in sealing performance. Development is in progress to increase the operating temperature range using strengthened (e.g., oxide-dispersion strengthened (ODS)) and refractory alloys. In laboratory tests, new nickel-based superalloy Rene 41 (Allvac, Inc.) seals have exhibited superior performance at 815 to 870 °C (1500 to 1600 °F) compared to standard Waspaloy seals with the same design. ODS alloys are being tested for temperatures above 870 °C (1600 °F).[15]

The U-Plex seal (fig. 11) is another self-energized static seal, similar to a multi-element E seal.[16] The E-seal is a single "folded" element. The U-Plex seal consists of two or more plies of materials nested together that act independently when the seal is compressed, as does a leaf spring, yet function as one under sealing pressure. It will accommodate 2.5 to 5 times more deformation than a single ply E-seal, is more compliant to surface irregularities, requires 1/3 the compression force, has enhanced high-cycle fatigue (HCF) resistance and has comparable leakage rates.

Added compliance (fig. 12(a)) is provided for combustor sealing of a large industrial gas turbine, where efficiency and emissions are key drivers for sealing advancements. Yet they bring with them increases in HCF, an aggressive thermal and pressure environment, and large thermomechanical transients and wear—all of which impact engine operations life (see also Appendix B). Flame stability and combustor emissions become highly dependent on uniformity of coolant and combustion air-fuel mixing. In turn uniform air supply depends on seals similar to the multiconvolution combustor liner seal, figure 12(a). Such a seal has been rig tested to withstand wear, oxidation, HCF, fretting, convolution pre-load loss, and leakage and has been proven in an integrated engine environment.[13] Field data are reported to corroborate the effectiveness of the seals in rig and scaled-up engine testing, but specific data are not provided.

As noted in the Introduction, it is important to remember that sealing interfaces and coatings are sacrificial. The cost of replacing or refurbishing a seal is minor relative to that of an engine combustor or turbine blade, for example.

B. Metallic Cloth Seals

Metallic and most nonmetallic fibers can be readily fabricated into a variety of configurations that are compliant and responsive to high-speed or lightly loaded systems, as projected by Hendricks.[17] Cloth seals are composite structures that make use of tightly woven typically metallic mesh. Most of the issues cited herein are applicable to the design of both metallic and nonmetallic cloth seals—providing proper attention is given to weaves and patterns, such as the multiple strands used in textile fabrics that require modifications for wear volume and contact area calculations. However, there are differences with the application of each, some of which are noted in Section III.C, "Cloth and Rope Seals."

For large gaps at interfaces with relative motion, rigid metal strips, feather seals, and "dog-bone" shaped strips have been the primary sealing methods. For applications with significant relative motion, these seals can either rock and rotate or jam against the slots in the adjacent components to be sealed. A lack of flexibility can result in poor sealing and excessive wear. Compliance can be attempted by reducing the thickness of the seal strips. But the use of thinner foil seals, as in aircraft engine applications, results in large stress levels and limited wear life in industrial turbine applications, which involve large interface gaps and demand much longer service life.† One approach to address seal compliance issues for large relative movements is the development of relatively low-cost, flexible cloth seals. Cloth seals are formed by combining thin sheet metals (shims) and layers of densely woven metal cloth. While shims prevent through leakage, and provide structural strength with flexibility, external cloth layers add sacrificial wear volume and seal thickness without adding significant stiffness. As illustrated in Fig. 12(b), a typical design requires simply wrapping a layer of metal cloth around thin flexible shims. The assembly is held together by a number of spot welds along the seal centerline.[10] Further leakage reduction can be achieved by a crimped design with exposed and contoured shim ends that enhance enwall sealing (Fig. 12(c))[89,19]. By maintaining contact with the slot surface, crimped shims better reduce the leakage flow. Demonstrated leakage reductions up to 30% have been achieved in combustors and 70% in nozzle segments. The flow savings achieved in nozzle-shroud cloth seal applications translate to large performance gains of up to a 0.50% output increase and 0.25% heat rate reduction in an industrial gas turbine.

†Thin foil seals are successfully used in aircraft engines where intersegment gaps and associated stress levels are small and overhaul intervals are much shorter than industrial applications, where typical expected life may extend to 64 000 hr.

Oxidation and wear resistance are the key attributes needed in the metallic-cloth fiber material. Likewise a structural shim must have high-temperature strength, and creep and fatigue resistance. A typical metallic-cloth fiber material is cobalt-based alloy Haynes 25, which is used for its superior high temperature wear resistance. For high-temperature applications, the cobalt-based superalloy Haynes 188 serves well as the shim material. While high mesh density is preferred for added flow resistance, a small fiber size required for increased density reduces oxidation life. Experience shows that Dutch twill weave with 30x250 fiber density per inch is the best cloth for sealing purposes.[20] In order to achieve improved wear resistance, the cloth weave should be oriented 45° to the dominant relative motion. Diagonal orientation also helps maintaining weave integrity if a local cut is incurred during operation.

Metallic-cloth seal design requires careful engineering to optimize flexibility while maintaining structural strength, resilience and robustness. A proper design process should include geometric analyses for engagement and jamming, finite element structural and stress analyses, rubbing wear tests, wear analyses, thermal flow analyses, subscale leakage performance tests, and analyses of leakage performance data. A detailed discussion on metallic-cloth seal design and analysis is given in Appendix B.

C. Cloth and Rope Seals

Rope or gasket seals can be used in various locations in turbomachinery. Table 1 lists the various materials being used. However, aircraft engine turbine inlet temperatures and industrial system temperatures continue to climb to meet aggressive cycle thermal efficiency goals. Advanced material systems including monolithic and composite ceramics, intermetallic alloys (i.e., nickel aluminide), and carbon-carbon composites are being explored to meet aggressive temperature, durability, and weight requirements. To incorporate these materials in the high-temperature locations of the system, designers must overcome materials issues, such as differences in thermal expansion rates and lack of material ductility.[21]

Designers are finding that one way to avoid cracking and buckling of the high-temperature brittle components rigidly mounted in their support structures is to allow relative motion between the primary and supporting components.[22] Often this joint occurs in a location where differential pressures exist, requiring high-temperature seals. These seals or packings must exhibit the following important properties: operate hot [≥ 705 °C (1300 °F)]; exhibit low leakage; resist mechanical scrubbing caused by differential thermal growth and acoustic loads; seal complex geometries; retain resilience after cycling; and support structural loads.

Braided rope seals can be made with a variety of materials and combinations, each having their own strengths and weaknesses. All ceramic designs consist of a ceramic fiber, uniaxial core, overbraided with ceramic sheath layers.[23] This design offers the potential for very high temperature 1150 °C (2100 °F) operation. However, researchers have determined that all ceramic seals are susceptible to the vibratory and acoustic loadings present in turbine engines. These seals can also be ejected from the seal gland due to dynamic loading.

To improve upon structural integrity, Steinetz and Adams[22] developed a hybrid braided rope seal design that consists of uniaxial ceramic core fibers overbraided with high-temperature superalloy wires. Tests have shown much greater resistance to abrasion and dynamic loadings.

Wires made of HS-188 material show promise to 870 °C (1600 °F) temperatures. This hybrid construction was used to seal the last-stage articulated turning vane of the F119 turbine engine. The seal limits flow of fan cooling air past the turning vane flow path (or powerstream)/fairing interface and also prevents backflow of potentially damaging high-temperature core air as shown in figure 13.[22]

Researchers at NASA Glenn continue to strive for higher operating temperature hybrid seals. Recent oxidation studies by Opila et al.,[24] showed that wires made from alumina forming scale base alloys (e.g., Plansee PM2000) could resist oxidation at temperatures to 1200 °C (2200 °F) for up to 70 hr. Tests showed that alumina-forming alloys with reactive element additions performed best at 1200 °C under all test conditions in the presence of oxygen, moisture and temperature cycling. These wire samples exhibited slow growing and adherent oxide scales. Dunlap et al.[25] provide experimental data of a 0.62-in. diameter rope seal that consisted of an Inconel X 750 spring tube, filled with Saffil insulation (Saffil Ltd.), and covered by two layers of Nextel 312 (3M Company) fabric wrap for operational temperatures to 815 °C (1500 °F). Steinetz and Dunlap[26] developed a braided carbon fiber thermal barrier that reduces solid rocket combustion gas leakage [3038 °C (5500 °F), in a nonoxidizing environment] and permits only relatively cool [< 93 °C (200 °F)] gas to reach the elastomeric O-ring seals. See also Section V.C, "Leaf and Wafer Seals."

In other related developments, Hendricks et al.[17] discussed the modeling and application of several types of brush seals including hairpin woven or wrapped (see hybrid seal), taconite, self-purging and buffer.

IV. Dynamic Seals

The inherent unsteady nature of turbomachines requires coupled solutions of powerstream and sealing interfaces. For example, rotor-stator-cavity interactions affect seal leakage and passage blockage factor. Nonsteady pressure distribution due to blade-vane interaction perturbs the leakage flows sucking them into the cavities near blade-vane coincidence and pumping them out near midcircumferential span position. These leakages lead to losses in component efficiency through injection of low momentum fluid into the powerstream, usually with more loss at hub; for smooth interfaces—for example, blade tip sealing—unsteady vortex formations dominate the losses. See also Appendix A for more details.

A. Tip Sealing

The flow field about the tip of the blades is illustrated in figures 14 and 15 for the compressor and turbine, respectively.[27] At the leading edge the flow is forced out and around the stagnation region, then joins with the primary leakage zone, and extends across the passage toward the low-pressure side and opposing the rotational velocity. These conditions are experimentally verified for tip clearance flows in the transonic compressor rotors and illustrated in figures 16 and 17.[28] Usually, the flow in transonic compressors is subsonic by the time it reaches the third or fourth stage.

Blade tip flows and ensuing vortex patterns lead to flow losses, instabilities and passage blockage. Without proper sealing, the flow field can be reversed, resulting in compressor surge, and possible fire at the inlet. Flow losses in static elements such as vanes in the compressor and turbine have different sealing requirements as cited later. A few of these dynamic interfaces for aero engine clearance control are illustrated in figure 2. More general flow details are found for example in Lakshminarayana[29] and for compressors in Copenhaver et al.,[30] Strazisar et al.,[31] Wellborn and Okiishi,[32] and for shrouded turbines in Bohn et al.[33]

B. Abradables

Early on, researchers recognized the need for abradable materials for blade tip and vane sealing, e.g., Ludwig,[4] Bill et al.,[9,34,35] Shimbob,[14] Stocker et al.,[36,37] and Mahler.[38] Schematics of three types of abradable materials with associated incursion types are illustrated in figure 18 for outer air-blade tip sealing interface in a compressor, for example. These types of materials usually differ from the platform or inner shroud-drum rotor interface sealing of the compressor as illustrated in figure 19.

As the name suggests, abradable seal materials are worn-in by the rotating blade during service. They are applied to the casing of compressors, and gas and steam turbines to decrease clearances to levels difficult to achieve by mechanical means. Abradable seals are gaining appeal in gas turbines as a relatively simple means to reduce gas-path clearances in both the compressor and turbine. They offer clearance reductions at relatively low cost and minor engineering implications for the service fleet. Abradable seals have been in use in aviation gas turbines since the late 1960s and early 1970s.[39] Although low energy costs, materials and long cycle time have in the past limited applications of abradable seals in land-based gas turbines, current operation demands enhanced heat rate and reduced costs. With increasing fuel prices and advances in materials to allow extended service periods, abradable seals are gaining popularity within the power generation industry.[40]

Without abradable seals, the cold clearances between blade or bucket tips and shrouds must be large enough to prevent significant contact during operation. Use of abradable seals allows the cold clearances to be reduced with the assurance that if contact occurs, the sacrificial part will be the abradable material on the stationary surface and not the blade or bucket tips. Also, abradable seals allow tighter clearances with common shroud or casing out-of-roundness and rotor misalignment.

1. Interface Rub

For properly designed abradables, if a rub occurs, the blade cuts into the sacrificial seal material with minimal distress to the blade. The abradable seal material mitigates blade wear while providing a durable interface that enhances engine efficiency. Controlled porosity shroud seal materials provide for low-energy material removal without damaging the rotating blade while mitigating leakage and enhancing seal life. Material release, porosity and structural strength can be controlled in both thermal-sprayed coatings and fibermetals. Filler materials are often used to resist energy input to the shroud seal, mitigate case clearance distortion, and also lubricate the wear interface. Worn material must be released to escape sliding contact wear of the blade tip (versus cutting action for an abradable) and plowing of the interface.[41] Asymmetric rubs generate hot spots that can develop into destructive seal drum instabilities. Such modes have destroyed engines and have been known to destroy aircraft with loss of life.

Many attempts have been made to study the wear mechanisms of abradable structures using conventional tribometers[42] or specially designed test rigs.[43,44] However due to the high relative speeds, >100 m/s (>330 ft/s), between the abradable seal and the rotating blade tip surface, the mechanisms of wear/cutting differ considerably from low speed tribology normally associated with machining operations. At high speeds, the removal/cutting of a thermal spray abradable coating is done by release of small particle debris, i.e., <0.1 mm (<0.004 in.). In contrast to conventional (low speed) cutting in machine tools, the particle debris released in abradable materials is ejected at the rear of the moving blade.[45] This, therefore partly sets the criteria for the design of such materials. It also sets a limiting design criterion for blade-tip thickness. Generally, a cutting element (blade-tip) thickness less than 1.3 mm (0.05 in) allows release of the particles from the coating. Thicker tips tend to entrap the loose particles between the blade and the abradable material. As a result, special considerations have to be given to the design of the materials to allow for the cutting mechanisms (for example, altering the base material particle morphology and size).

Certain abradable materials rely more on densification (compaction) of the structure than on particle debris removal.[46] Material compaction limits the functional depth of the abradable material since the compacted material will increase the wear of the rotating blade tips as the porosity is reduced. These types of seal materials include some of the thermal spray coatings and porous metal fiber structures (fiber metals). Fiber metals can be designed and constructed with varying fiber sizes and densities to alter their tribological behavior.[47,48]

2. Interface Materials

There are different approaches used for a material to be abradable. Some have porosity built in so the material wears away when rubbed by the blade tip. Some of these have a solid lubricant embedded to aid the wear process. Other abradables, such as honeycomb and fiber metal, deform at high speeds and the cell walls rupture. For honeycomb, rotor wear is most pronounced at the brazed web where cell thickness doubles. Borel et al.[46] mapped incursion velocity as a function of tangential velocity as shown in figure 20. These parameters delineate regions of adhesive wear, melting wear, smearing, cutting and adhesive titanium transfer from blade to interface. Abradable seals are generally classified according to their temperature capability,[49] but can also be characterized by method of application as shown in table 2.[41]

a) Low temperature abradable seals.—For thermally sprayed abradable coatings, different classes of coating materials behave tribologically differently. Traditionally, most of the powder metals available for low temperature applications, that is., <400 °C (<750 °F), are aluminum-silicon based. To make them abradable, a second phase is added.[49] This phase is usually a polymeric material or a release agent and is often called a solid lubricant. The role of the second phase in aluminum-silicon based abradable material is primarily to promote crack initiation within the structure. The size, morphology, quantity, and material of the second phase determine the wear mechanisms and abradability of the seal coating under various tribological conditions. The wear map in figure 20 is for an aluminum-silicon-polyester coating. The dominant wear mechanisms are different for various combinations of blade-tip velocity and incursion rate when rubbed by a 3 mm (0.12 in) thick titanium blade at ambient temperature. The arrows indicate the movement of wear-mechanism boundaries when a stiffer polymer than polyester is used as the second phase.

Low temperature abradables (generally epoxy materials) are used for fan tip sealing. Engine manufacturers' philosophy regarding fan rub strips is engine dependent. For example, the PW4090 uses a filled-honeycomb configuration, shown in figure 21(a). The uneven rub, caused by in-flight maneuvers, can become, relatively speaking, quite deep, (tens of mils) which is difficult to tell from the photo. The PW4000 and PW2000 have very similar labyrinth style rub-strips figure 21(b). On the other hand, the CFM56 engine uses a smooth surface, which gets repotted during overhaul, and yet is usually not refurbished unless considerable damage has been incurred.

b) Mid-temperature abradable seals.—For temperature applications up to 760 °C (1400 °F), Ni- or Co-based alloy powders are commonly used as the basis of the abradable seal matrix. Other phases are added to the base metal powder to make the material abradable. These added phases are polymeric materials that are fugitive elements to generate coating porosity and act as release agents.[44,50]

Figure 22 displays a wear map of a mid-temperature coating system abraded at 500 °C (930 °F) using titanium blades. The map shows the wear mechanism domains vs. blade-tip velocity and incursion rate. The arrows indicate the movement of the wear regime boundaries as the polyester level increases. As polyester content and thus porosity increases, cutting becomes increasingly predominant mechanism over the entire range of the speeds and incursion rates. However, increasing porosity has a negative effect on coating cohesive strength and erosion properties.

Fiber metals are a type of mid-temperature abradable. Like other abradables, abradability and erosion resistance present conflicting design demands, as illustrated in figure 23, and provide the seal designer with some flexibility.[47] Chappel et al.[47,48] tested different fiber metals against other abradables for high and low-speed abradability and erosion (see table 3). The materials were then ranked per their performance (table 4). Results showed that the high-strength fiber metal had the best performance overall with the highest abradability and lowest erosion.

*c) **High temperature abradable seals**.*—For operating temperatures above 760 °C (1400 °F), common practice is to use porous ceramics as the abradable material. The most widely used material is YSZ, which is usually mixed with a fugitive polymeric phase. There are a number of important considerations regarding porous ceramic abradable materials. To achieve an acceptable abradability, the cutting element/blade generally has to be reinforced with hard abrasive grits. Choosing these grits and processes to apply them has been the subject of numerous research activities. There are a number of patents that deal with this aspect of ceramic abradable materials.[51–55] Abrasive grits considered include cubic boron nitride (cBN), silicon carbide, aluminum oxide and zirconium oxide. Published data suggest that cBN particles of a given size range tend to be the best abrasive medium against YSZ porous ceramics.[51,54] Cubic boron nitride poses a high hardness (second to diamond) and a high sublimation temperature, >2980 °C (>5400 °F), which makes it an ideal candidate to abrade ceramic abradable materials. But cBN's relatively low oxidation temperature, ~850 °C (~1560 °F), allows it to function for only a limited time. This has prompted the use of other abrasives such SiC.[52,55] Despite successful functionality of SiC against YSZ, SiC has been met with limited enthusiasm. SiC requires a diffusion barrier to prevent its reaction with transition metals at elevated temperatures.[56] This adds to the complexity and the cost of the abrasive system.

The ceramic abradable coating microstructure and its porosity are other essential considerations. Clearly, porosity increases the abradability of the coating. However, YSZ is strongly susceptible to high angle erosion because of its brittle nature,[57] and adding porosity makes it prone to low angle erosion. Thermally sprayed porous YSZ coatings show different tribological behavior when compared to metallic abradable materials. They tend to show a strong influence of blade-tip velocity on abradability[45] (see fig. 24). Abradability tends to improve with increasing blade-tip velocity. On the other hand, porous YSZ coatings show less dependency on incursion rate. They tend to have poor abradability at very low incursion rates, <0.005 mm/s (< 0.2 mils/s), thus requiring blade-tip treatments.

An example application of a high temperature abradable has been reported where the bill-of-material (BOM) first-stage turbine gas path shroud seals were coated with a porosity controlled plasma-sprayed partially-stabilized zirconia (PSZ) ceramic as shown in figure 25.[58] The coating was a 1 mm (0.040 in.) layer of ZrO_2-$8Y_2O_3$ over a 1 mm (0.040 in.) NiCoCrAl-based bond coat onto a Haynes 25 substrate. Characteristic "mudflat" cracking of the ceramic occurred at the blade interface, but back-side seal temperature reductions over BOM-seals of 78 °C (140 °F) were measured, with gas path temperatures estimated over 1205 °C (2200 °F).

3. Designing Abradable Materials for Turbomachinery

Because abradable seals are low strength structures that wear without damaging the mating blade tips, they are also susceptible to gas and solid particle erosion. Abradable structures intended for use in harsh temperatures occurring in gas turbines can also be prone to oxidation because of the inherent material porosity. These conflicting properties need to be accounted for in designing abradable seals. Abradable seals then have to be considered as a complete tribological system that incorporates 1) Relative motions and depth of cut—blade-tip speed and incursion rate; 2) Environment temperature, fluid medium and contaminants; 3) Cutting element geometry and material—blade-tip thickness, shrouded or unshrouded blades; 4) Counter element—abradable seal material and structure. Manufacturing processes as well as microstructural consistency of abradable seals can have a profound effect on their properties.[47,59]

Another issue to consider in designing of abradables for compressors is the large changes in thermal environment and the fact that titanium fires are not contained. Therefore, rubbing must release particulate matter without engendering a fire or debris impacting downstream components. Also in compressors, an abradable can be combined with intentional grooving to enhance stall margins, yet clearance control or fluid injection may be better methods of controlling stall margin.

Considering all the above design elements makes the abradable system quite unique, that is, designed to suit the particular application. Thus, despite the availability of many off-the-shelf materials, abradable seals have to be modified or redesigned in most applications to meet the design constraints. More extensive lists of references on abradable seals and their use have been published elsewhere.[45,47,49]

C. Labyrinth Seals

Labyrinth seals and their sealing principles are commonplace in turbomachinery and come in a variety of configurations. The most used configurations are straight, interlocking, slanted, stepped and combinations, (fig. 26).[60] By their nature labyrinth seals, usually mounted on the rotor, are clearance seals that can rub against their shroud interface, such as abradables and honeycomb (fig. 27).[61] They permit controlled leakages by dissipation of flow energy through a series of sequential aperture cavities (as sequential sharp edge orifices) with minimum heat rise and torque. The speed and pressure at which they operate is only limited by their structural design.

Principle design parameters include: clearance and throttle (tooth or knife) and cavity geometry and tooth number (fig. 28).[62] The clearance is set by aerothermo-mechanical conditions that preclude contact with the shroud allowing for radial and axial excursions. The throttle tip is as thin as structurally feasible to mitigate heat propagation through the throttle-body into the shaft with a sharp leading edge (as an orifice) and is the primary flow restrictor. The angle at which the flow approaches the throttle is usually 90° but slant throttles, into the flow, are more effective seals (Borda inlets, [C_f = 0.5, where C_f is the flow coefficient (actual flow/ideal isentropic flow)], are more restrictive than orifice inlets, [C_f = 0.63]). One advantage of 90° throttles (C_f = 0.63) is the ability to seal flow reversals equally well; slant throttles are less effective handling flow reversals (C_f = 0.8 to 0.9). The cavity geometry is nearly 1:1 with axial spacing greater than six times the clearance and often shaped to enhance flow dissipation through generation of vortices. A relation between the number of knife-cavity modules and leakage for developed cavity flows is given by

$$G_r/G_{r1} = N^{-0.4}$$

where G_r = mass flux and G_{r1} is the mass flux through a single throttle and N the number of throttles or cavity-throttle modules,[63] for gas throttles only, see Egli[64] for an equivalent relation (Egli's interest was steam, yet applicable to gases in general). Conditions relating the sharpness of the tooth to the ability to restrict flows are given by Mahler[38] (fig. 29), and more recently explored by computational fluid dynamics (CFD).[65]

Labyrinth seals are good in restricting the flow but do not respond well to dynamics and often lead to turbomachine instabilities. These problems have been addressed by several investigators starting with Thomas[66] and Alford.[67] They recognized that the dynamic forces drove instabilities and heuristically determined stable operating configurations (fig. 30). Benckert and Wachter,[69] Childs et al.,[70] and Muszynska[71] addressed the root causes and introduced the swirl brake at the seal inlet to mitigate the circumferential velocity component within the cavities, (fig. 31). More recently, circumferential flow blocks and flow slots have been introduced to mitigate the circumferential velocity component, (figs. 32 and 33).[12, 724]

Labyrinth seals have a lengthy history of proven reliability with robust operation and developed technology and are well suited for abradable interfaces. Their tendency to engender instabilities can be controlled by swirl brakes or intra-cavity slots or blocks and drum dampers. Nearly all turbomachines rely on labyrinth seals or labyrinth sealing principles (Egli,[64] Trutnovsky,[73] Stocker et al.,[36] and Stocker[37]). In general, nearly all sealing applications rely heavily on the essential features of sharp-edge flow restrictors [e.g., the aspirating seal (see section V.E, "Oil Brush Seals") has a labyrinth tooth and the face-sealing dam,[74] figure 34(a) and (b), and the inlet throttle confining flows to the honeycomb land (fig. 35)].

D. Brush Seals

As described by Ferguson,[75] the brush seal is the first simple, practical alternative to the finned labyrinth seal that offers extensive performance improvements. Benefits of brush seals over labyrinth seals include 1) Reduced leakage compared to labyrinth seals (upwards of 50 percent possible); 2) Accommodate shaft excursions due to stop/start operations and other transient conditions. Labyrinth seals often incur permanent clearance increases under such conditions, degrading seal and machine performance; 3) Require significantly less axial space than labyrinth seal; and 4) More stable leakage characteristics over long operating periods.

Brush seals have matured significantly over the past 20 years. Typical operating conditions of state-of-the-art brush seals are shown in table 5.‡

Brush seal construction is deceptively simple, requiring the well ordered layering or tufting of fine-diameter bristles into a dense pack that compensates for circumferential differences between inside and outside diameters, (figs. 36 and 37). This pack is sandwiched and welded between a backing ring (downstream side) and sideplate (upstream side), then stress relieved to insure

‡Data available at http://www.fluidsciences.perkinelmer.com/turbomachinery.

stability and flatness. The weld on the seal outer diameter is machined to form a close-tolerance outer diameter-sealing surface to fit into a suitable housing. The wire bristles protrude radially inward (shaft-rotor) or outward (drum-rotor) and are machined to fit the mating rotor, with slight interference. Brush seal interferences (preload) must be properly selected to prevent catastrophic overheating of the rotor and excessive rotor thermal growths.

To accommodate anticipated radial shaft movements, the bristles must bend. To allow the bristles to bend without buckling, the wires are oriented at an angle (typically 45° to 55°) to a radial line through the rotor. The bristles are canted in the direction of rotor rotation. The bristle lay angle also facilitates seal installation, due to the slight interference between the bristle pack and the rotor. The backing ring provides structural support to the otherwise flexible bristles and assists the seal in limiting leakage. To minimize brush seal hysteresis caused by brush bristle binding on the back plate, new features have been added to the backing ring. These include reliefs of various forms. An example design is shown in figure 36 and includes the recessed pocket and seal dam. The recessed pocket assists with pressure balancing of the seal and the relatively small contact area at the seal dam minimizes friction allowing the bristles to follow the speed-dependent shaft growths. The bristle free-radial-length and packing pattern are selected to accommodate radial shaft movements while operating within the wire's elastic range at temperature. A number of brush seal manufacturers[§] include some form of flow deflector (e.g., see flexi-front plate in figs. 36 and 37) on the high pressure side of the wire bristles. This element

---

[§] Data available online at http://www.crossmanufacturing.com.

aids in mitigating the radial pressure closing loads (e.g., sometimes known as "pressure closing") caused by air-forces urging the bristles against the shaft. This element can also aid in reducing installation damage, bristle flutter in highly turbulent flow fields, and FOD.

Brush seals, initially developed for aero-gas turbines, have also been used in industrial gas and steam turbines since the 1990s. Design similitude, analysis and modeling of brush and woven seals were established earlier in the works of Flower[76] and Hendricks et al.[17] Within in the confines of this section we are only able to address a few sealing types, their locations, and their material constraints. For further details, see Hendricks and coworkers[27,77,78] and NASA Conference Publications.[79,80] An extensive summary of brush seal research and development work through 1995 has been published[84,85] and updated in a more recent summary.[40]

1. Brush Seal Design Considerations

To properly design and specify brush seals for an application, many design factors must be considered and traded-off. Comprehensive brush seal design algorithms have been proposed by Chupp et al.,[41] Dinc et al.,[83] Hendricks et al.,[17] and Holle and Krishan.[84] An iterative process must be followed to satisfy seal basic geometry, stress, thermal (especially during transient rub conditions), leakage, and life constraints to arrive at an acceptable design. Many of the characteristics that must be considered and understood for a successful brush seal design are given here:[83] pressure capability, seal upstream protection, frequency, seal high- and low-cycle fatigue (HCF, LCF) analysis, seal leakage, seal oxidation, seal stiffness, seal creep, seal blow-down (e.g., pressure closing effect), seal wear, bristle-tip forces and pressure stiffening effect, solid particle erosion, seal heat generation, reverse rotation, bristle-tip temperature, seal life/long term considerations, rotor dynamics, performance predictions, rotor thermal stability, oil sealing, secondary flow and cavity flow (including swirl flow), and shaft considerations (e.g., coating, etc.). Design criteria are required for each of the different potential failure modes including stress, fatigue life, creep life, wear life, oxidation life, amongst others. Several important designs parameters are discussed next.

a) Material selection.—Materials in rubbing contact in brush seal installations must have sufficient wear resistance to satisfy engine durability requirements. A proper material selection requires knowledge of the rotor and seal materials and their interactions. In addition to good wear characteristics, the seal material must have acceptable creep and oxidation properties.

Metallic bristles: Brush seal wire bristles range in diameter from 0.071-mm (0.0028-in.) (for low pressures) to 0.15-mm (0.006-in.) (for high pressures). The most commonly used material for brush seals is the cobalt-based alloy Haynes 25 based on its good wear and oxidation characteristics. Brush seals are generally run against a smooth, hard-face coating to minimize shaft wear and the chances of wear-induced cracks from affecting the structural integrity of the rotor. The usual coatings selected for aircraft applications are ceramic, including chromium carbide and aluminum oxide. Selecting the correct mating wire and shaft surface finish for a given application can reduce frictional heating and extend seal life through reduced oxidation and wear. There is no general requirement for coating industrial gas and steam turbine rotor surfaces where the rotor thicknesses are much greater than aircraft applications.

Nonmetallic bristles: High-speed turbine designers have long wondered if brush seals could replace labyrinth seals in bearing sump locations. Brush seals would mitigate traditional labyrinth seal clearance opening and corresponding increased leakage. Issues slowing early application of brush seals in these locations included: coking (carburization of oil particles at excessively high temperatures), metallic particle damage of precision rolling element bearings, and potential for fires. Development efforts have found success in applying aramid bristles for certain bearing sump locations.[85,86] Advantages of the aramid bristles include: stable properties up to 150 °C (300 °F) operating temperatures, negligible amount of shrinkage and moisture absorption, lower wear than Haynes 25 up to 150 °C, lower leakage (due to smaller 12-μm-diameter fibers), and resistance to coking.[85] Based on laboratory demonstration, the aramid fiber seals were installed in a GE 7EA frame (#1) inlet bearing sealing location. Preliminary field data showed that the nonmetallic brush seal maintained a higher pressure difference between the air and bearing drain cavities and enhanced the effectiveness of the sealing system allowing less oil particles to migrate out of the bearing. More will be said about these types of nonmetallic brush seals in Section V.E, "Oil Brush Seals" and Appendix C.

b) Seal fence height.—A key design issue is the required radial gap (fence height) between the backing ring and the rotor surface. Following detailed secondary flow, heat transfer, and mechanical analyses, fence height is determined by the relative transient growth characteristics of the rotor vs. the stator and rotordynamic considerations. This backing ring gap is designed to avoid contact with the rotor surface during any operating condition with an assumed set of dimensional variations. Consequently, the successful design of an effective brush seal hinges on a thorough knowledge of the turbine behavior, operating conditions, and design of surrounding parts.

c) Brush pack considerations.—Depending on required sealing pressure differentials and life, wire bristle diameters change.[87] Better load and wear properties are found with larger bristle diameters. Bristle pack widths also vary depending on application: the higher the pressure differential, the greater the pack width. Higher-pressure applications require bristle packs with higher axial stiffness to prevent the bristles from blowing under the backing ring. Dinc et al.[83] have developed brush seals that have operated at air pressures up to 2.76 MPa (400 psid) in a single stage. Brush seals have been made in very large diameters. Large brush seals, especially for ground power applications are often made segmented to allow easy assembly and disassembly, especially on machines where the shaft stays in place during refurbishment.

d) Seal stress/pressure capability.—Pressure capacity is another important brush seal design parameter. The overall pressure drop establishes the seal bristle diameter, bristle density, and the number of brush seals in series. In a bristle pack, all bristles are essentially cantilever beams held at the pinch point by a front plate and supported by the back plate. From a loading point of view, the bristles can be separated into two regions (see fig. 36). The lower part, fence region, between the rotor surface and the back plate inner diameter (ID), and the upper part from the back plate ID to the bristle pinch point. The innermost radial portion carries the main pressure load and is the main source of the seal stress.[88] In addition to the mean bending stress, contact stress at the bristle-back plate interface must be considered. Furthermore, bristle stress is a very strong function of the fence height set by the expected relative radial movement of the rotor and seal.

Figure 38 shows a diagram illustrating design considerations for seal stress and deflection analysis, and includes a list of the controllable and noncontrollable design parameters. As a word of caution, care must be taken in using multiple brush configurations as pressure drop capability becomes more nonlinear with fluid compressibility and most of the pressure drop or bristle pressure loading is carried by the downstream brush.

e) Heat generation/bristle tip temperature.—As the brush seal bristles rub against the rotor surface, frictional heat is created that must be dissipated through convection and conduction and is quite similar to the classic Blok problem,[89] where extensive heating occurs at the sliding interface. Brush seal frictional heating was addressed by Hendricks et al.17.[90] and modeled as fin in crossflow with a heat source at the tip by Dogu and Aksit.[91] If the seal is not properly designed, this heating can lead to premature bristle loss, or worst, the rotor/seal operation could become thermally unstable. The latter condition occurs when the rotor grows radially into the stator increasing the frictional heating, leading to additional rotor growth, until the rotor rubs the seal backing plate resulting in component failure. In some turbine designs, brush seals are often assembled with a clearance to preclude excessive interference and heating during thermal and speed transients. These mechanical design issues significantly affect the range of feasible applications for brush seals. Many of these issues have been addressed by Dinc et al.[83] and Soditus.[92]

f) Seal leakage.—Leakage characterization of brush seals typically consists of a series of tests at varying levels of bristle-to-rotor interference or clearance, as shown in figures 39 and 40. Static (nonrotating) tests are run to get an approximate level of seal leakage and pressure capability. They are followed by dynamic (rotating) tests to provide a more accurate simulation of seal behavior. Rotating tests also reveal rotor dynamics effects, an important consideration for steam turbine rotors and turbomachines in general, that can be sensitive to radial rubs due to nonuniform heat generation.

Proctor and Delgado studied the effects of speed [up to 365 m/s (1200 ft/s)], temperature [up to 650 °C (1200 °F)] and pressure [up to 0.52 MPa (75 psid)] on brush seal and finger seal leakage and power loss.[93] They determined that leakage generally decreased with increasing speed. Leakage decreases somewhat with increasing surface speed since circumferential flow is enhanced and the rotor diameter increases; changes in diameter causes both a decrease in the effective seal clearance and an increase in contact stresses (important in wear and surface heating).

g) Other Considerations.—If not properly considered, brush seals can exhibit three other phenomena deserving some discussion. These include seal "hysteresis," "bristle stiffening," and "pressure closing." As described by Short et al.[87] and Basu et al.,[94] after the rotor moves into the bristle pack (due to radial excursions or thermal growths), the displaced bristles do not immediately recover against the frictional forces between them and the backing ring. As a result, a significant leakage increase (more than double) was observed following rotor movement.[94] This leakage hysteresis exists until after the pressure load is removed (e.g., after the engine is shut down). Furthermore if the bristle pack is not properly designed, the seal can exhibit a considerable stiffening effect with application of pressure. This phenomenon results from interbristle friction loads making it more difficult for the brush bristles to flex during shaft excursions. Air leaking through the seal also exerts a radially inward force on the bristles, resulting in what has been termed "pressure closing" or bristle "blow-down." This extra contact load, especially on the upstream side of the brush, affects the life of the seal (upstream bristles are worn in either a scalloped or coned configuration) and higher interface contact pressure. By measuring baseline seal leakage in a line-to-line (zero clearance) assembly configuration, bristle blowdown for varying loads of assembly clearance can be inferred from leakage data (see fig. 40).

2. Brush Seal Flow Modeling

Brush seal flow modeling is complicated by several factors unique to porous structures, in that the leakage depends on the seal porosity, which depends on the pressure drop across the seal. Flow through the seal travels perpendicular to the brush pack, through the annulus formed between the backing ring bore and the shaft diameter. The flow is directed radially inward towards the shaft as it flows around individual bristles and collides with the bristles downstream in adjacent rows of the pack and finally between the bristle tips and the shaft.

A flow model proposed by Holle et al.,[95] uses a single parameter, effective brush thickness, to correlate the flows through the seal.¶

¶To use the Holle et al. flow model, corrections for density ρ must be included: For laminar flow

$(Re_v \leq 100)$, $G_{max}^2 = \dfrac{2g_c}{180(144)}(\Delta P)\rho \, Re_v \, \dfrac{HD_{v,m}}{B} \left(\dfrac{S_{T,m}}{HD_{v,m}} \right)^{0.4} \left(\dfrac{S_{T,m}}{S'_{L,m}} \right)^{0.6}$, and from the *characteristic* point 1 the postlaminar friction factor for the interpolation in the *transition flow* region (Re$_v$ > 100 and Re$_b$ < 5000) $f_{KL_1} = \dfrac{g_c}{2(144)} \dfrac{(\Delta P)\rho}{G_{max_1}^2} \dfrac{HD'_m}{B}$.

See Holle et al. for symbol definitions.

Variation in seal porosity with pressure difference is accounted for by normalizing the varying brush thicknesses by a minimum or ideal brush thickness. Maximum seal flow rates are computed by using an iterative procedure that has converged when the difference in successive iterations for the flow rate is less than a preset tolerance.

Flow models proposed by Hendricks et al.[17,90,96] are based on a bulk average flow through the porous media. These models account for brush porosity, bristle loading and deformation, brush geometry parameters and multiple flow paths. Flow through a brush configuration is simulated using an electrical analog with driving potential (pressure drop), current (mass flow), and resistance (flow losses, friction and momentum) as the key variables. All of the above mentioned brush flow models require some empirical data to establish correlation constants. Once the constants are established, the models can predict brush seal flow reasonably well.

A number of researchers have applied numerical techniques to model brush seal flows and bristle pressure loadings.[97–100] Though these models are more complex, they permit a more detailed investigation of the subtleties of flow and stresses within the brush pack.

3. Applications

a) Aero gas turbine engines.—Brush seals are seeing extensive service in both commercial and military turbine engines. Lower leakage brush seals permit better management of cavity flows and significant reductions in specific fuel consumption when compared to competing labyrinth seals. Allison Engines (now Rolls Royce) has implemented brush seals for the Saab 2000, Cesna Citation-X, and V-22 Osprey. General Electric has implemented a number of brush seals in the balance piston region of the GE90 engine for the Boeing 777 aircraft. Pratt & Whitney has entered revenue service with brush seals in three locations on the PW1468 for Airbus aircraft and on the PW4084 for the Boeing 777 aircraft.[101]

b) Ground-based turbine engines.—Brush seals are being retrofitted into ground-based turbines both individually and combined with labyrinth seals to greatly improve turbine power output and heat rate.[40,83,102–106] Dinc et al., report that incorporating brush seals in a GE Frame 7EA turbine in the high pressure packing location increased output by 1.0 percent and decreased heat rate by 0.5 percent.[83] Figure 41 is a photo of a representative brush seal taken during a routine inspection. The seal is in good condition after nearly three years of operation (~22,000 hr). To date, more than 200 brush seals have been installed in GE industrial gas turbines in the compressor discharge high-pressure packing (HPP), middle bearing, and turbine interstage locations. Field data and experience from these installations have validated the brush seal design technology. Using brush seals in the interstage location resulted in similar improvements. Brush seals have proven effective for service lives of up to 40,000 hr.[83]

E. Face Seals

Labyrinth seals are less impacted by FOD-debris than other types of seals, yet also pass that debris to other components such as bearing cavities. One of the major functions of face and buffer sealing is to preclude debris from entering the bearing or gear-box oil yet an equally important function is to prevent oil vapors from leaking into the wheel-space and from entering the cabin air stream. Debris in the bearing or gear-box oil can radically shorten life and oil-vapor in the wheel space can cause fire or explosions. Oil vapors in the cabin are unacceptable to the consumer-traveler.

Face seals are classified as mechanical seals. They are pressure balanced contact or self-acting seals. The key components are the primary ring (stator) or nosepiece, seat or runner (rotor), spring or bellows preloader assembly, garter or retainer springs, secondary seal and housing (figs. 42 and 43).[108,109] There is a wealth of information on the experimental data for, design of, and application of mechanical seals in the literature from technical publications to books; For example, see Ludwig[4] and Lebeck.[110]

For the face seal, the geometry of the ring or nosepiece becomes critical. For successful face sealing, the forces due to system pressure, sealing dam pressure and the spring or bellows must be properly balanced and stable over a range in operating parameters (pressure, temperature, surface speed) (fig. 44).[111]

Contact seals wear and are generally limited to surface speeds less than 76 m/s (250 ft/s). To mitigate the wear, prolonging life and decreased leakage Ludwig[112] and Dini[113] promoted the self-acting Rayleigh step and spiral groove seal, (figs. 45 to 47). A labyrinth seal or a simple projection representing a single throttle is used for presealing to control excessive leakage should the dam of the face seal "pop" open; for example, the labyrinth preseal as is illustrated in figure 45 (and aspirating seal of section V.E). Spiral groove (fig. 47), slot and T-grooving (bidirectional) are more commonly used than Rayleigh steps to provide more lift at less cost to manufacture.

Self-acting seals permit tighter clearances and better control of the sealing dam geometry as sealing pressure drops are increased, providing lower leakage. Figure 48 provides a comparison of the leakage rates between labyrinth, face-contact and self-acting seals. While self-acting face sealing greatly reduces leakage, surface speeds are generally limited to less than 213 m/s (700 ft/s), but nearly triple the limits of contact face sealing 61 to 91 m/s (200 to 300 ft/s).

F. Oil Seals

Gas turbine shaft seals are used to restrict leakage from a region of gas at high pressure to a region of gas at low pressure. A common use of mechanical seals is to restrict gas leakage into bearing sumps. Oil sealing of bearing compartments of turbomachines is difficult. A key is to prevent the oil side of the seal from becoming flooded. Still, oil-fog and oil-vapor leakage can occur by diffusion of oil due to concentration gradients and oil transport due to vortical flows within the rotating labyrinth-cavities (crude distillation columns). Bearing sumps contain an oil-gas mixture at near-ambient pressure, and a minimal amount of gas leakage through the seal helps prevent oil leakage and maintains a minimum sump pressure necessary for proper scavenging. Bearing sumps in the HPT are usually the most difficult to seal because the pressure and temperatures surrounding the sump can be near compressor discharge conditions.

1. Radial Face Seals

Conventional rubbing-contact seals (shaft-riding and radial face types) are also used to seal bearing sumps. Because of their high wear rates, shaft-riding and circumferential seals (fig. 42) have been limited to pressures less than 0.69 MPa (100 psi); and successful operation has been reported at a sealed pressure of 0.58 MPa (85 psi), a gas temperature of 370 °C (700 °F), and sliding speed of 73 m/s (240 ft/s).[108]

2. Ring Seals

The ring seal, as described by Whitlock[114] and Brown,[109] is essentially an expanding or contracting piston ring. The expanding design is simpler and is illustrated in figure 49. Other designs that can be grouped in the ring seal family include the circumferential segmented ring seal and the floating or controlled-clearance ring, as described by Ludwig.[4] The material requirements for these seals are essentially the same as those for the expanding ring seal. The ring seals are carbon and they seal radially against the inside diameter of the stationary cylindrical surface as well as axially against the faces of the adjacent metal seal seats (fig. 49). The metal seal seats are fixed to, and rotate with, the shaft. The sealing closing force is provided by a combination of spring forces and gas pressures. Ring seals are employed where there is a large relative axial movement due to thermal mismatch between the shaft and the stationary structure. Ring seals are limited to operation at air pressure drops and sliding speeds considerably lower than those allowed for face seals. However, they can be used to gas temperature levels in the same range as for positive-contact face seals, approximately to 480 °C

(900 °F). Generally, a minimum pressure differential of 14 kPa (2 psid) must be maintained to prevent oil leakage from the bearing compartment.

Carbon ring and face sealing of the sumps described by Ludwig,[4] Whitlock,[114] and Brown[109] are fairly standard. Boyd et al.[115] have investigated a hybrid ceramic shaft seal which is comprised of a segmented carbon ring with lifting features as the outer or housing ring and a silicon-nitride tilt-support arched rub runner mounted on a metal flex beam as the inner ring (fig. 50). The flex beam added sufficient damping for stability and no oil see page was seen at idle speed down to pressure differentials of 0.7kPa (0.1 psia), air to oil.

3. Materials

Selecting the correct materials for a given seal application is crucial to ensuring desired performance and durability. Seal components for which material selection is important from a tribological standpoint are the stationary nosepiece (or primary seal ring) and the mating ring (or seal seat), which is the rotating element. Brown[109] described the properties considered ideal for the primary seal ring as shown here: 1) mechanical—high modulus of elasticity, high tensile strength, low coefficient of friction, excellent wear characteristics and hardness, self-lubrication; 2) thermal—low coefficient of expansion, high thermal conductivity, thermal shock resistance, and thermal stability; 3) chemical—corrosion resistance, good wetability; and
4) miscellaneous—dimensional stability, good machinability, and low cost and readily available.

Because of its high ranking in terms of satisfying these properties, carbon graphite is used extensively for one of the mating faces in rubbing contact shaft seals. However, in spite of its excellent properties, the carbon material must be treated in order for it to satisfy the operational requirements of sealing applications in the main rotor bearing compartment of jet engines.

Seal failures are driven by thermal gradient fatigue or axial and radial thermal expansions during maximum power excursions. Bearing compartment carbon seals will fail from the heat generated in frictional rub. Excessive face wear occurs during transients and, as mentioned, labyrinth seals can allow oil transport out of the seal and oil contamination by the environment (moisture, sand, etc.).[116]

G. Buffer Sealing

Public awareness of environmental hazards, well-publicized effect of hazardous leakages (Three Mile Island, Challenger), and a general concern for the environment, have precipitated emissions limits that drive the design requirements for sealing applications. Of paramount concern are the types of seals, barrier fluids, and the necessity of thin lubricating films and stable turbomachine operation to minimize leakages and material losses generated by rubbing contacts.[107]

A zero-leakage seal is an oxymoron. Industrial practice is to introduce a buffer fluid between ambient seals and those seals confining the operational fluid (fig. 51) with proper disposal of the buffered fluid mixture.[112,114] A second example is for shaft sealing as shown in figure 52 where buffer fluids are introduced. In the case of oil sumps, the buffered mixture is vented to the hot gas exhaust stream and is presumed to be consumed. Within the nuclear industry, this becomes a containment problem where waste storage now becomes an issue. In the case of rocket engines, the use of buffering or inerting fluids (e.g., helium) is commonplace to separate fuel and oxidizer-rich environments for example in the Space Shuttle main engine (SSME) turbomachinery.

H. Rim Sealing and Disk Cavity Flows

Turbomachine blade-vane interactions engender unsteady seal and cavity flows in multiply connected cavities with conjugate heat transfer and rotordynamics. A comprehensive review of seals-secondary flow system developments are documented by Hendricks et al.[117,118] and NASA Seals Code and Secondary Flow Systems Development publications.[80]

Unsteady flows perturb both the power and the secondary flow streams.[2] A T1 turbine (first stage of the HPT) can have 76 blades and 46 stators all interacting with unsteady loadings (fig. 53).[119] Cavity ingestion of rapidly pulsating hot gases induce cavity heating, increases disk temperature, which in turn limits disk life and can compromise engine safety. Proper sealing confines these gases to the blade platform regions.

Rotordynamic issues further complicate rim seal and interface seal designs. These issues are addressed in: Thomas,[66] Alford,[67,120,121] Benckert and Wachter,[69] NASA Conference Publications,[79] Abbott,[68] von Pragenau,[122] Vance,[123] Childs,[124] Muszynska,[71] Bently and Hatch,[125] Hendricks,[118] and Temis.[126]

Cavity and sealing interface requirements differ between industrial and aero-turbomachines. Major differences include split casings and through bolted disks, and compressors and turbines with common drive shafts for industrial machines vs. cylindrical casings and drum rotors on multiple spools for aero machines. Figure 53 shows a typical aero multistage turbine cavity section. Several experimental studies have been reported that consider both simplified and complex disk cavity configurations (e.g., Chen;[127] Chew et al.[128,129] Graber et al.,[130] and Johnson et al.[131,132]). Cavity sealing is complex and has a significant effect on component and engine performance and life. However, several analytical and numerical tools are available to help guide the designer, experimenter and field engineer in addressing these challenges (see appendix A).

V. Advanced Seal Designs

A. Finger Seal

The finger seal is a relatively new seal technology developed for air-to-air sealing for secondary flow control and gas path sealing in gas turbine engines.[133-135] It can easily be used in any machinery to minimize airflow along a rotating or nonrotating shaft. Measured finger seal air leakage is 1/3 to 1/2 of conventional labyrinth seals. Finger seals are compliant contact seals. The power loss is similar to that of brush seals.[136] It is reported that the cost of finger seals are estimated to be 40 to 50 percent of the cost to produce brush seals.

The finger seal is comprised of a stack of several precisely machined sheet stock elements that are riveted together near the seal outer diameter as shown in figure 54. The outer elements of the stack, called the forward and aft coverplates, are annular rings. Behind the forward coverplate is a forward spacer, then a stack of finger elements, the aft spacer and then the aft coverplate. The forward spacer is an annular ring with assembly holes and radial slots around the seal inner diameter that align with feed-thru holes for pressure balancing. The finger elements are fundamentally an annular ring with a series of cuts around the seal inner diameter to create slender curved beams or fingers with an elongated contact pad at the tip. Each finger element has a series of holes near the outer diameter that are spaced such that when adjacent finger elements are alternately indexed to the holes, the spaces between the fingers of one element are covered by the fingers of the adjacent element. Some of the holes create a flow path for high pressure upstream of the seal to reach the pressure balance cavity formed between the last finger element, the aft spacer and seal dam, and the aft coverplate. The aft spacer consists of two concentric, annular rings. One is like the forward spacer. The second is smaller with an inner diameter the same as the aft coverplate and forms the seal dam. It is connected to the outer annular ring by a series of radial spokes.

The fingers provide the compliance in this seal and act as cantilever beams, flexing away from the rotor during centrifugal or thermal growth of the rotor or during rotordynamic deflections. The pressure balance cavity reduces the axial load reacted by the seal dam and hence minimizes the frictional forces that would cause the fingers to stick to the seal dam and cause hysteresis in the finger seal leakage performance. In this seal there are two leakage paths. One is thru (around and under) the fingers at the seal/rotor interface. The other is a radial flow across the seal dam. When a pressure differential exists across the seal the fingers tend to move radially inward towards the rotor. Test results confirm this pressure closing effect. The pressure closing effect is largely due to the pressure gradient under the finger contact pads. The bulk of the radial pressure loads on the curved beam of the finger balance out to a zero net load. Ideally, one would design finger seals to have a line-to-line fit during operation. However, most applications involve a range of operating conditions and seal-to-rotor fits and clearances change due to different coefficients of thermal expansion, centrifugal rotor growth, pressure closing effects, and dynamics of the rotor. Depending on the requirements of the application it may be desirable to start with an interference-fit at build and allow the seal to wear in or it may be desirable to have a clearance between the seal and rotor at build and allow the gap to close up.

Finger seals are contacting seals and wear of the finger contact pad is expected. Life is dependent on the materials selected and operating conditions. Arora et al.[134] reported the seal and rotor were in excellent condition after a 120 hr endurance test. Testing of Haynes-25 fingers against a Cr_3C_2 coated rotor resulted in a wear track on the rotor 0.0064mm (0.00025 in.) deep. The finger seal wore quickly to a near line-to-line fit with the rotor.[135]

B. Noncontacting Finger Seal

Altering the geometry of the basic finger seal concept, Braun et al.[137] and Proctor and Steinetz[138] developed a new seal that combines the features of a self-acting shaft seal lift pad as an extension of the downstream finger with an overlapping row of noncontacting upstream fingers (figs. 55 and 56). These lift pads are in very close proximity to the rotor outer diameter so that hydrodynamic lift can be generated during shaft rotation. The seal geometry is designed such that the hydrostatic pressure between the downstream lift pad and the shaft is slightly greater than that above the pad. Depending on the application and operating conditions, the designer may choose to integrate hydrodynamic lift geometries on either the rotor or pads (e.g., taper, pockets, steps, etc.) to further increase lift-off forces. The overlapping fingers reduce the axial and radial flows along the compliant fingers that allow for radial motion of the shaft-seal interface. These seals respond to both radial and axial shaft perturbations and some degree of misalignment with minimal hysteresis. This technology is still being developed, but some experimental and analytical work has shown its feasibility.[139] It is expected that noncontacting finger seals will have leakage performance approximately 20 percent higher than a contact finger seal, which is still significantly better than conventional labyrinth seals, but have near infinite life since they won't rub against the rotor, except very briefly at start and stop. Both the finger seal and noncontacting finger seal are in the development stage. To the authors' knowledge, neither seal has been tested in an engine.

C. Leaf and Wafer Seals

Continued efforts in high-temperature rope and wafer sealing are directed improving braid wire, wafer and preload-spring materials.[114] Silcon nitride (Si_3N_4) is favored for seal wafers (fig. 57 (b)). Worn-in wafer seals provide 1 to 1½ orders of magnitude less leakage than rope seals, yet consideration must be given to the interface application: leakage, film cooling, or thermal barrier. The wafer seal (fig. 57) can be preloaded by several methods: for example, bellows, coil springs, or canted coil springs. Canted coil springs offer near-constant loading characteristics over a range of compression. Coil springs are reported to maintain strength to 1200 °C (2190 °F); refractory metals as Mo-47.5Re (molybdenum rhenium), Mo-0.5Ti-0.08Zr (TZM), and coatings of iridium and rhodium (oxidation resistant coatings used in some rocket thruster engine nozzle applications, (for example, see Reed and Schneider[140]) are projected to afford superior high temperature applications.[141]

The leaf seal as described by Flower[142] and Nakane et al.[143] (fig. 57) is an adaptation of the wafer seal advanced by Steinetz and Sirocky[144] with principles of operation delineated by Hendricks et al.,[26,90] Steinetz and Hendricks[111] and Nakane et al.[143] The leaf and wafer seals have similar encapsulation but differ in root attachment and moments of inertia or cross section. The stacked leaves (or wafers) are relatively free to move in the radial direction and are deformable along the length or circumference providing a compliant restrained two-dimensional motion as opposed to the brush-seal-bristle, which deforms in three dimensions.

Nakane et al.[143] reported leakage performance of a leaf seal at less than 1/3 that of an equivalent four stage, 0.5 mm gap labyrinth seal geometry when run back to back on the same test rotor, with little wear at the smooth coated rotor interface. Variations in front and back plate gaps are used to control lift of the leaves based on pressure drop and rotor speed. Modeling of a leaf or wafer sealing is similar (fig. 58) where pressure balances, thickness, length, inclination, housing gaps and attachment points all require proper treatment for flexure and gap spacing. The latter are difficult to assess and flow coefficients are most often determined experimentally. With that in mind, computations provided by Nakane et al.[143] are in good agreement with experimental data and CFD results, with little interface wear. Leaf-seal, leakage, endurance and reliability are being evaluated in a M501G in-house industrial gas turbine. Both leaf and wafer seals provide for radial, circumferential and axial shaft motions. Even though the wafer seal is depicted with a low mobility wall, this restraint is not required. Evaluation of the rotordynamic stability coefficients for these seals is still required. Stiffness and damping is similar to brush configuration modeling but with an altered cross section. These seals will not respond well to grooved or rough surfaces, which are alleviated when lift off occurs.

An alternate form of the leaf seal has been advanced by Gardner et al.[145] The seal is comprised of overlapping shaft-riding leaves (thin-metallic-sheets), which extend from the sealing inlet about the shaft, forming shaft-riding fingers (fig. 59). The cantilevered inner-leaves form lifting-pads that are overlapped by outer-leaves that appear as cantilevered J-springs. The outer-leaf, which sees system pressure, seals the cavity and permits compliant radial excursions. It is similar in configuration to a film riding compliant foil bearing. This design also allows for hydrostatic operation, namely when pressure is applied, the interface deforms to maintain an operating film, even without rotation. The overlapping shingled elements float on the fluid film providing excellent sealing with virtually no wear, yet can be somewhat limited by their ability to handle large system pressure and radial excursions without damaging the leaves. Gardner[145] reported hydrostatic and hydrodynamic performance results for a 121 mm (4.75-in.) diameter seal. The pressure-balanced seal permitted hydrostatic liftoff independent of rotor speed with total liftoff and no shaft torque with a 0.48 MPa (70 psid) pressure differential. The seal produced a torque of 0.028 N-m (0.25 in-lb) at 1200 rpm with a 0.42MPa (60 psid) pressure differential and a radial displacement of 0.23 mm (0.009-in.). When compared to typical industrial and an aerospace four-tooth labyrinth seal, with a 0.152 mm (0.006-in.) clearance and 1995 vintage-brush seal configurations, the leaf seal had leakage characteristics 1/4 that of the industrial labyrinth at design conditions and about 1/3 that of the brush (fig. 60).[146]

D. Hybrid Brush Seals

Justak[147,148] combined a brush seal with tilt pad-bearing concepts to eliminate bristle wear. He introduced two designs: (1) the bristles are attached to the pads, and (2) the pads are supported via beam elements and the bristle tips remain in contact with the outer surface of the pads (see figs. 61 and 62 respectively). The brush seal stiffness is based on the design flow conditions and rotational speed. In the spring beam design, the beam elements are sized to allow radial movement and restrict axial displacements. In either design, seal leakage is controlled by the brush and lift off by the pads, but leakage is more constant than a conventional brush seal and can accommodate reverse rotation with no bristle wear. Testing with rotor offsets up to 0.51 mm (0.020 in.) showed no wear or temperature rise and 1/3 less torque to maintain speed when compared to a conventional brush seal.[149]

Shapiro[150] combined shaft, face, and brush sealing concepts to form a film-riding seal with "L"-shaped annular segments (pads) that are preloaded onto the shaft via the brush seal and held in place by a garter spring (fig. 63). The brush seal is separated axially from the brim of the cylindrical segments via a spring. This design allows the axial position of the brush seal to alter the preload and hence stiffness of the cylindrical seal segments. Analytical studies show low leakage, compliance, stability, and low wear during seal operation, yet development is required.

The Hybrid Floating Brush Seal (HFBS) (fig. 64) combines a brush seal and a film riding face seal that allows both axial and radial excursions mitigating interface problems of friction, heat and wear.[149] The brush seal forms the primary seal, rotates with the shaft, while floating against a secondary face seal that acts as a thrust bearing. The HFBS relies on a high interference fit or preload. Major advantages include elimination of wear between rotating and stationary components, improved sealing performance, and the ability to handle large axial and radial shaft excursions while maintaining sealing integrity. For a 71 mm (2.8 in.) diameter HFBS that allowed up to 6.4 mm (0.25-in.) axial travel, experimental results showed 1/6 the leakage of a standard brush seal at the same operating pressure ratios and rotational speed and an order of magnitude less than numerical predictions of a standard labyrinth seal (see fig. 65).[153–155]

E. Oil Brush Seals

Brush seals are being considered for oil and oil mist sealing applications in turbomachinery (see fig. 66).[153] Brush seals perform very well under rotor transients, owing to the inherent compliance of bristles. Application areas include oil and bearing sumps in gas turbines and aircraft engines, and as buffer seals in hydrogen cooled industrial generator applications; they have been suggested for vehicle applications.[17] Typically oil-sump applications require oil mist sealing such as at the front bearing to prevent oil mist ingestion into the compressor.[152,155] Yet brush seals are primarily contact seals, and in addition to the leakage rate, oil temperature rise and coking become major concerns. The inclined approach at the tip of individual bristles creates small hydrodynamic bearing surfaces at the rotor-bristle interface. The amount of lift determines the critical balance between oil temperature rise and leakage rate. Typically nonmetallic fibers are used in oil brush seals because of concerns of metal particle generation near bearings. Bristle shrinkage, wear resistance, inertness, and moisture absorption rates are major concerns when selecting bristle fiber material.[155] Although oil applications of brush seals are rather new, their use in gas turbine front bearing applications have proven successful. Field tests have shown leakage reduction gains and have demonstrated durability of these seals in field operation.[154,155] Further discussion on fiber material selection and seal analysis is provided in Appendix C.

F. Aspirating Seals

An aspirating seal is a hydrostatic face seal with a narrow gap to control the leakage flow and a labyrinth tooth to control leakage flow at elevated clearances (fig. 34(a)) and forms a high-pressure cavity for the diffused-injected flow in the engine cavity at operating conditions (fig. 34(b)).[74]

Conventional labyrinth seals are typically designed with a seal/rotor radial clearance that increases proportionally with diameter. Aspirating face seals are noncontacting seals that are designed to establish an equilibrium position within close proximity of the rotor surface regardless of the seal diameter.[156–161] Aspirating seals have a potentially significant performance advantage over conventional labyrinth seals, particularly at large diameters. In addition, these seals are inherently not prone to wear, owing to their noncontacting nature, so their performance is not expected to degrade over time. Figure 34(a) and (b) shows a cross section of the seal design, which is enhanced by the presence of a flow deflector on the rotor face.

During operation the aspirating face seal performs as a hydrostatic gas bearing. The gas bearing on the sealing face provides a thin, stiff air film at the interface; as the clearance between the seal face and the rotor decreases, the opening force of the gas bearing increases. The seal face geometry is designed to give an operating clearance of 0.038 mm to 0.076 mm (0.0015 to 0.003 in.). In operation, the spring forces play a minor role; thus the seal is free to follow the rotor on axial excursions. Tolerances between the primary face seal ring and the housing allow the ring to tilt relative to the housing; the seal can thus follow the rotor even if there is an angle between the face and the rotor, such as during a maneuver or due to rotor runout.

When there is an insufficient pressure drop across the seal to maintain an adequate film thickness (such as during startup and shutdown, periods when conventional face seals would touch down), springs retract the seal from the rotor face (fig. 34(a)). This ensures that the seal never contacts the rotor, thus providing for long seal life. In the retracted position, the pressure drop across the seal occurs at the aspirator tooth. During startup, as the pressure drop rises, the pressure balance across the primary face seal ring "aspirates" the seal to a closed position (fig. 34(b)). The labyrinth tooth provides the required pressure drop to close the seal as well as a fail-safe seal in case of failure of the aspirating face seal.

Tests have been conducted to evaluate prototype performance under a variety of conditions that the seal might be subjected to in an aircraft engine application, including cases of rotor runout and seal/rotor tilt.[156–162] The tests were executed on a full-scale 36-in. (0.91-m) diameter rotary test rig. Analyses were performed using three-dimensional CFD in order to validate test data and to establish the seal design. The full-scale tests demonstrated that with the flow isolation tip, a hydrostatic film forms at the air bearing resulting in a seal/rotor clearance of 0.025 to 0.038 mm (0.001 to 0.0015 in.)., with correspondingly low leakage rates. The seal performs effectively with rotor runouts as great as 0.25 mm (0.010 in.) total indicator reading, and the seal was able to accommodate the expected angular misalignment (tilt) of 0.27°.

G. Microdimple

Laser surface texturing, termed microdimples, (fig. 67)[163] is a further extension of the damper bearing and seal-bearing work established by von Pragenau[164,165] and extended by Yu and Childs[166] who found that a hole-area to surface-area ratio of 0.69 is a more effective seal than honeycomb. For the microdimpled seal (fig. 67) where diameter is 125 ± 5 μm (4900 ± 200 μ-in.) and depth 2.5 ± 0.5 μm (98 ± 20 μ-in.) with 0.2 μm (8 μ-in.) surface finish, the hole/surface area ratio is 0.3 indicating some potential for improvement. Diamond-like graphite or antifouling coatings were not used, but afford potential improvements.

H. Wave Interfaces

Etsion[167] also developed a wave pumping face seal, modified by Young and Lebeck[168] to a wave interface; these concepts have been combined and further improved by Flaherty et al.[169] and are considered more debris tolerant, (fig. 68).

I. Seal-Bearing

Munson et al.[7] describe and provide operations data for room temperature testing of a seal-bearing concept. The basic concept was advanced by von Pragenau.[122,165] In Munson's et al. seal configuration, the foil thrust bearing is combined with a mating flat interface to make a device called a foil face seal (fig. 69). Multiple wave bump foils support the interface foils. With pressure drop and rotation, this interface gives rise to a compliant hydrodynamic film-riding face seal. For a 20,000 lb (89-kN)-thrust class engine, with this technology, an estimated mission fuel burn reduction of 1.85 percent for a fixed engine and "rubber-airframe" and 3.17 percent for both engine and airframe being "rubber" was reported. (Here "rubber" refers to allowing for design parameter changes.)

J. Compliant Foil Seal

Expanding upon Gardner's leaf seal concept,[145] Salehi and Heshmat[170] proposed an extension of their foil bearing work as a seal (fig. 70).

Forming an inexpensive, close tolerance, bell-mouth smooth interface foil, similar to a nozzle-inlet, is not an easy task, yet can be accomplished by flow form or shear form spinning.** However, Salehi and Heshmat[170] and Heshmat[171] chose to form the bellmouth-nozzle inlet by cutting radial relief slots to account for the difference between inner and outer circumference (diameters) and bending the tabs to form the bell-mouth or "L-shaped" foil section. The resulting foil is then attached to the housing at one end opposing rotation. The slot relief spacing is dependent on stresses in the bend radius, foil thickness and seal diameter to prevent significant "pleating" of the seal-interface foil. The resulting near-smooth compliant, noncontacting foil interface rides on a fluid film, typically < 0.0127 mm (< 500 µ-in.) thickness. The "L-shaped" section provides blockage for the bump foil opening and must be carefully contoured at the shoulder and spring-pressed against the support structure forming the necessary secondary seal. Overlapping leaves can assist to minimize these slot leakages yet it is a difficult area to seal. The

**Data available from Franjo Metal Spinning, at http://www.franjometal.com/metal-spinning/flow-forming.html#.

interface foil (or foils) are in turn supported on a series of bump foils that provide variable stiffness in response to radial shaft excursions. The foils are usually coated with a solid lubricant to minimize startup and shutdown interface contact wear while permitting radial leaf sliding at the inlet.

Such seals are low leakage fluid film devices that are capable of operating at high surface velocities, temperature, and pressure loadings limited by the foil materials used in the construction, for example, 365 m/s (1200 fps) and 600 °C (1100 °F).

K. Deposits Control

For turbines operating in high salt environments, Nalotov[172] introduces strategically placed holes within the turbine shroud ring to provide equalization of circumferential pressures within the labyrinth interface. The concept is shown in figure 71(a) and (b). Figure 71(a) is a sketch representing the cross section of a nozzle and shrouded turbine stage where salt and metal oxides have built up on the shroud. Figure 71(b) shows the location of the discharge holes. The pressure equalization induces stability inside the shroud chamber, which allows for reduced shroud seal clearance. The flow through the shroud ring produces an obstacle effect to prevent deposits from building up and hence reduces blade passage blockage by accumulated salts. The concept has the net effects of increasing turbine engine efficiency as well as service life.

L. Active Clearance Control (ACC)

Most modern turbomachines have variable geometry controls. The compressor, for example, uses inlet guide vanes that are rotated to enhance efficiency at off-design conditions (usually designed to handle takeoff (TO) and set for cruise). In large aero-engines, case clearance control is used in the turbine. Today's larger commercial engines control HPT blade tip clearances by impinging fan air on the outer case flanges. Systems such as those shown in figure 72(a) and (b) scoop air from the fan by-pass duct to cool the outer case flanges, reducing the case diameter and hence shroud clearance. Other engines use a mixture of fan and compressor air to achieve finer HPT tip clearance control. Because these thermal systems are relatively slow, they cannot be used during transient events such as TO and re-acceleration. As such they are generally scheduled for operation during cruise conditions. Lattime and Steinetz[173] are developing fast-response systems that utilize clearance measurement feedback control, enabling true active clearance control at engine startup and throughout the flight envelope (fig. 73). Active clearance control is not usually used in the compressor, rather the efficiency is enhanced through varying the vane angle and vortex control through fluid injection. Without these systems several points in efficiency are lost. Yet such close control is not without potential blade-shroud and vane-rotor rubbing where the cited efficiency may be lost.

VI. Life and Limitations

System design conditions for seal controlled component cooling are driven by compliance to regulatory agencies, reliability and safety standards.[174] The mean time between failures (MTBF) is highly dependent on the thermal loading (fig. 74) and the aero-engine and flight operations profile (figs. 75 and 76).

For engine component life modeling, the time τ_i and the life L_i under thermomechanical load from the environmental temperature (fig. 77) and flight envelope profile (figs. 75 and 76) are used to determine the cumulative loss of component life according to the linear damage rule of Palmgren, Langer, and Miner.[175]

Zaretsky et al.[176] applied Weibull-based life and reliability analysis to rotating engine structures. The NASA E^3 engine design data served as the basis for the analysis.[2,177] When limits are placed on stress, temperature, and time for a component's design, the criterion that will define the component's life and thus the engine's life will be either high-cycle or low-cycle fatigue.

Knowing the cumulative statistical distribution (Weibull function) of each engine component is a prerequisite to accurately predicting the life and reliability of an entire engine. The columns in table 6 shows how some of the hot section component lives correlate to aero-engine maintenance practices without and with refurbishment, respectively. That is, it can be reasonably anticipated that at one of these time intervals, 5 percent of the engines in service will have been removed for repair or refurbishment for a cause.

Within the open literature there is a dearth of data for seals and their functional life and for basic materials. The classic approach is deterministic and assumes that full and certain knowledge exists for the service conditions and the material strength. This means that specific equations that define sealing conditions are coupled with experience-based safety factors, yet it is well known that variations with loading can have a significant effect on component reliability. The Weibull-based analysis addresses these issues. Yet, until a sealing database is established, the MTBF will continue to be based on field experience.

Summary

Turbine engine cycle efficiency, operational life and systems stability depend on effective clearance control. Designers have put renewed attention on clearance control, as it is often the most cost-effective method to enhance system performance. Advanced concepts and proper material selection continue to play important roles in maintaining interface clearances to enable the system to meet design goals. No one sealing geometry or material is satisfactory for general use. Each interface must be assessed in terms of its operational requirements. Insufficient clearances limit coolant flows, cause interface rubbing, and engender turbomachine instabilities and system failures. Excessive clearances lead to losses in cycle efficiency, flow instabilities and hot gas ingestion into disk cavities. Hot gas ingestion in the turbine cavities reduces critical disk life and in the bearing sump location engenders bearing and materials failures. Reingestion of flow along the compressor drum interface causes unnecessary blockage and can lead to compressor stall.

Materials play a major role in maintaining interface clearances. Abradable materials for the fan are usually polymers; for the LPC compressor, ambient to 400 °C (750 °F): fiber metals and AlSi + filler can be used, but for the midrange LPC and HPC, ambient to 760 °C (1400 °F): Ni or Co base can be used (titanium blade fire protection limits); and if the blades are Ni-based superalloys, NiCrAl-Bentonite might be a choice. In the HPT, 760 °C (1400 °F) to 1150 °C (2100 °F), yttria-stabilized zirconia (YSZ) with controlled porosity and cBN or preferably SiC blade tip abrasive grits can be used, depending on how hot the engine runs; in general, air plasma spray thermal barrier coatings (APS-TBC's) are used in the combustor, and for some engines, first-stage vanes (nozzles) and second-stage blades of the HPT. Electron beam plasma vapor deposition (EB-PVD) TBC's are used on the HPT T1 or first-stage blades, some second-stage blades and some first-stage vanes (nozzles). TBC's are not commonly used in the LPT due to lower heat flux and are less effective in decreasing component temperature. APS ceramics are also used on shroud seals (blade outer air seals) where they function as both a thermal barrier for the metallic shroud and abradable seal.

Component life and reliability are closely coupled with the duty cycle. But as energy demands (and emissions regulations) necessitate more time-responsive-controlled engines, the industrial and aero-engine duty cycles become similar.

Acknowledgments

Sealing in turbomachinery has been the focus of numerous development efforts. Many of the developers have been cited in this chapter. The authors would like to especially acknowledge contributors to this review: Margaret Proctor (finger seals), Norm Turnquist (aspirating seals), Saim Dinc and Mehmet Demiroglu (brush seals), Stephen Stone and Greg Moore (metallic static seals), Farshad Ghasripoor (abradables), and Glenn Holle (an extensive review). We also wish to thank our sponsoring organizations for the time and resources to prepare this sealing review.

Appendix A.—Further Discussion on Rim Sealing and Disk Cavity Flows

Coupling of the powerstream, seals and cavity flows is a necessary aspect of multistage compressor and turbine design. Athavale et al.[178,179] have reported detailed descriptions of these tools and representative simulations. (See also Janus and coworkers).[180,181] These works include the gas-path flows through the stages (blades and vanes) interacting with those under the platform and within the cavity as well as the sealing interface (shaft to platform), (fig. A1).[182] Interstage-labyrinth-rim seal interface design goals are to keep leakage small, reduce windage and blockage, mitigate ingestion and reintroduction of leakage flows.

Teramachi et al.[183] investigated turbine rim interface sealing (figs. A1 and A2), providing data and some CFD results on four rim seal configurations: (0) T-on rotor, (1) T-on rotor with overlap T-on stator, (2) T-on stator with overlap T-on rotor; and (3) fish mouth on rotor with overlap T-on stator. Dummy stators were introduced, but there were no blades on the rotor. Carbon dioxide concentration measurements (similar to the work of Johnson[132] (Graber et al.[130])) defined seal effectiveness in terms of the ratio of purge gas to ingested gas.

Figure A3 shows the seal effectiveness of these configurations, where flow coefficient $C_w = Q/\nu b$ and $\text{Re}_m = Vb/\nu$, where b is the cavity outer radius, V is the mean flow speed, Q is the purge flow rate, and ν is the kinematic viscosity. Configuration (3) is the least affected by changes in overlap and configuration (0) the most; configuration (2) is quite sensitive to overlap. The high effectiveness of configuration (3) is related to the buffer cavity between the two rotor seal teeth with a stator tooth between the rotor teeth. The lowest effectiveness of configuration (1) is due to the large clearance gap, although the ingestion is nearly zero. CFD results show the gap recirculation zone where powerstream gas is ingested at on-pitch positions and ejected at mid-pitch positions (i.e., flow ingestion at the vane leading edge partially returns in the mid-pitch region mixing with the purge air; see also Wellborn and Okiishi[30,184]).

Wellborn and Okiishi[32,184] investigated the effect of leakage in a four-stage, LPC with blading design based on the NASA E^3 engine. Seal leakages did not affect upstream stages but did progressively degrade performance of the downstream stages. For each 1 percent change in clearance/span ratio the pressure rise penalty was nearly 3 percent with a 1 percent drop in efficiency. Hall and Delaney[185–187] simulated the low-speed axial compressor (LSAC) experiments with Adamczyk's analysis package.[188] They also completed sensitivity studies but did not address the effects on rotordynamics.

Heidegger et al.[62] presented three-dimensional solutions of the interaction between the powerstream and seal cavity flow in a typical multistage compressor (fig. A4). Using the Allison/NASA-developed ADPAC code, they performed a parametric study on a three-tooth labyrinth seal/cavity configuration and a sensitivity study to various sealing parameters. Their study shows that the leakage flow out of the seal cavities can affect the powerstream significantly, mainly by altering the inlet flow near the stator blade root area, and can potentially affect the performance of the overall compressor (fig. A5).

Bohn et al.[33] investigated the interaction between the powersteam end-wall cavities and sealing clearances in a two-stage, shrouded-blade turbine. Three cases were considered: A—engineering design, where blade-sealing cavities interactions are neglected, B—end-wall sealing gaps adjusted to near-zero clearance, and C—end-wall radial gaps approximately 0.8 mm (fig. A6), representing both shroud and hub configurations.

Periodic flow ingestion and circumferential fluid shuttling depends on relative blade-vane circumferential position ($\Delta\theta$) as noted in fig. A5 and implied in Figs. 16 and 17 (for compressors); time-dependent unshrouded turbine solutions are discussed in the next section. Similar shroud and hub flow patterns are found at the shrouded turbine interfaces, configurations B and C. At the shrouded interface, flows are sucked into the cavities near $\Delta\varphi \approx 0$ shuttled circumferentially and ejected near $\Delta\varphi \approx$ mid-span, fig. A7. Hub interface flow patterns are illustrated in fig. A8. In both cases, the injection of low momentum fluid into the powerstream increases passage flow blockage and decreases component polytrophic efficiency; however the effect is more detrimental at the hub than at the shroud. These computational results were verified experimentally for Bohn et al.[189] who also investigated unshrouded blading configurations.

Feiereisen et al.[190] completed an experimental study of the primary and secondary flows in a turbine rig. It represents a first attempt at understanding this interaction and at generating data for validation. CFD techniques provide detailed flow field information on complex cavity shapes that cannot be treated with analytical methods (e.g., Athavale et al.,[191,192] Chew,[130] Virr et al.,[193] and Ho et al.[194]).

Ho et al.[194] and Athavale et al.[195] in studying the Allison 501D turbine found that ingested fluid could work its way well into the disk space, even though purge fluid flows were substantial. Figures A9 to A11 show the calculated interaction between powerstream and secondary flows. Without conjugate heat transfer, the calculations would not match the Allison 501D turbine design-data. It is most important at powerstream interfaces where the thermal gradient from the platform to the hub is significant. These aero-thermomechanical loads can drastically affect disk and engine life.

The coupled codes (SCISEAL and MS–TURBO)[65] have been applied to several experimental test rig data sets showing conditions under which ingested flow can be controlled. The configuration (figs. A12 and A13), is a 30° pi-sector with four vanes (stators) and five blades (rotors) simulating the stator/rotor set (48/58) with a three-tooth labyrinth seal and overlap rim seals.[190] Athavale et al.[65] found that a recirculation zone in the rim seal was present at the lower purge flow rate but was absent at the higher purge flow rate. The recirculation allows some gas ingestion into the rim seal area. This gas can then travel inside the cavity by both diffusion and convection, (fig. A14). Two important observations can be made:

(1) the interface velocities show a tangential component that is lower than the rotor speed. This slow fluid alters the angle of attack near the roots of the rotor blades and can cause loss of power (turbine) and stall (compressor), and (2) the rotor blades have the expected upstream pressure rise, which affects the flow in the rim seal and the cavity (enhances ingestion), although this disturbance is rather small.

Smout et al.[196] in a CFD analysis of rim sealing cite some collaborative efforts involved in investigating rotating cavity ventilation, bearing cavity purge and cooling, pressure balance, and sealing rotor/stator gaps. The turbine slinger determines the preswirl of cooling air entering the HPT blades. Controlling preswirl in power and secondary flow streams becomes very important for rotordynamics and power on demand cycling. For preswirl analysis and control methods see

Thomas,[66] Benckert and Wachter,[69] NASA Conference Publications,[79] von Pragenau,[121] Childs,[124] Muszynska,[71] Bently and Hatch,[125] and Hendricks.[118]

Appendix B.—Further Discussion of Metallic Cloth Seals

In this Appendix attention is confined to metallic cloth seals, yet as stated in the text, if proper properties are used, the analysis can be used for other weaves like ceramics and the like. However, if multiple strands are used instead of wires—such as fabrics in textiles—some modifications may be needed in wear volume and contact area calculations.

Cloth seals are used at the junctions between the components in combustor and turbine sections (figs 3 (b) and B1. High temperatures in these sections require intersegment gaps to accommodate large thermal expansions as well as to aid in cooling the parts. Large varying gaps with relative motion of the mating parts require proper sealing. A seal should be sufficiently thin and compliant to accommodate large misalignments. On the other hand, long inspection intervals and large vibration levels, which are typical of large industrial engines, require thick seal sections to provide sacrificial wear volume. A lack of flexibility can result in poor sealing and excessive wear. Cloth seals, formed by combining thin sheet metals (shims) and layers of densely woven metal cloth, address the compliance and sealing performance issues. Decoupling structural and wear-related elements of the seal design allows the independent optimization of sheet metal for high-temperature strength, while the cloth material and weave can be optimized for maximum wear and oxidation resistance.

1. Seal Construction

(a) **Interstage seals:**

Typically, nozzle and shroud intersegment junctions have slots for seal strips. As illustrated in figs. 3, 12, and B2, cloth seal designs for such applications require a simple wrapping of a layer of cloth around thin flexible shims. Further leakage reduction can be achieved by a crimped design where shims are crimped over the edge of the cloth layers.[10,17,18–20,199] These composite cloth seal strips replace commonly used nozzle-shroud intersegment metallic splines that are inserted in the deep slots in the mating parts. Properly designed crimps reduce the leakage flow very effectively. However, care should be taken to ensure sufficient flow remains, as required for slot and seal cooling.

(b) **Combustor seals:**

Combustion dynamics and excessive thermal misalignments make combustor sealing more challenging than the nozzle-shroud intersegment applications. In gas turbines with can-annular combustion systems, the combustor seals are used to seal the gap between the can transition duct (TD) and the first-stage nozzles (FSNs). A typical sealing junction involves two TDs (cans) and multiple FSN segments. Large axial offsets and relative skew misalignments between neighboring cans are quite common. As shown in fig. B3, these junctions are typically sealed using formed metal strips designed to take relative axial and radial motion by sliding in grooves machined in the TD and FSN,[198] fig B1. However, the FSN is made of segments that experience relative misalignments causing the seal to stick in the FSN slot. Jamming the seal on the FSN side results in wear of the seals by the TD because of the relative dynamic motion. Heavy wear on the seal and in the TD slots is commonplace. Seal failure can cause occasional forced power outages. Combustor cloth seals have addressed this need for flexibility at the TD-FSN junctions.[198] As illustrated in fig. B4, TD-FSN cloth seals utilize a radial lip formed by a flexible cloth-shim assembly. Cloth seals also incorporate an interference fit providing a uniform seal-slot contact under any condition, thereby providing reduced leakage. When the seal is jammed in the FSN slot, relative vibratory motion is absorbed by flexing of the cloth assembly rather than the wearing on the rigid seal frame.

2. Materials Selection

When selecting materials for a cloth seal design, oxidation and wear resistance are the key attributes needed in the cloth fiber material. Likewise, structural shim must have high-temperature strength and oxidation resistance. Creep and fatigue properties are also important for the shim. A typical cloth fiber material is Haynes 25 (also referred to as L605), which is used for its superior high-temperature wear resistance. However, for applications beyond 750 °C (~1400 °F), the increased oxidation rate may reduce life drastically. For higher temperatures, Haynes 188 can be considered. Haynes 188 has excellent oxidation resistance, albeit it with higher wear rates, requiring typical engineering tradeoffs, see Sections IV.B.2, "Interface Materials," and IV.B.3, Designing Abradable Materials for Turbomachinery."

Inconel X-750 is the shim material of the choice for applications less than 600 °C (~1100 °F). It is a precipitation-hardened, high-strength, and fatigue-resistant spring material. Typically, in this temperature range Inconel X-750 is used for combustor seals. Haynes 188 is used for nozzles and shrouds where seal temperatures run higher. The combustor seals run cooler: first, because coolant air is colder at the combustor section then gets hotter as it moves through the turbine section resulting in less efficient cooling and second, because the combustor cloth seals are placed in the cooler transition piece aft frame slot, which is embedded in cold compressor discharge air.

3. Weave Selection

Proper cloth selection involves cloth weave, mesh density and orientation. The most common metal cloth constructions can be grouped as plain, twill, Dutch twill and stranded weaves. Cloth seal weave selection is based on key parameters pertaining to wear and leakage performance as well as mesh integrity. Dutch twill weave combines unequal wire diameters with staggered and alternating passes (see fig B5). It makes the optimal construction for cloth seal designs as it combines benefits of a high-density mesh with relatively large fiber diameters. Dutch twill cloth also offers an interlocking construction, which results in higher mesh integrity during local cuts. Although there are tighter weave types like micron weave, these require small wires to achieve high density. High-temperature gas turbine sealing applications require oxidation resistance, making larger wire diameters more beneficial.[199]

Cloth selection requires leakage and wear testing of various cloth samples. As illustrated in fig B6, there are four relevant orientations for a cloth weave: along the warp wires (1A), along the shute wires (1B), diagonal (2), and normal to the cloth surface (3). The warp wires are the main wires running the length of woven cloth. The shute wires run perpendicular to the warp wires or across the cloth as woven. These are sometimes referred to as "fill" or "weft" wires. Three different leakage tests are performed in the plane of the cloth. Cloth samples are tested parallel and orthogonal to the warp direction of the weave as well as diagonally to the weave direction.

An additional leakage test is performed normal to the cloth surface, measuring the cross flow. Apart from the cross flow, leakage in other orientations depends more on mesh density than the orientation. Experience shows that diagonal orientation yields the best wear performance.[20] Diagonal orientation also helps maintain weave integrity if a local cut is incurred during operation. Finally, using cloth in diagonal orientation enhances flexibility by allowing the mesh to distort rather than pull on warp or shute wires during deformations.

4. Design Considerations

Cloth seal design requires careful engineering to optimize flexibility while maintaining structural strength and robustness. The inherent decoupling of structural and wear-related components allows better control over the design. Before designing a seal, the operating conditions (or flight design envelope) need to be established. Some of the major tasks for a proper cloth seal design include:

- Optimizing seal dimensions to prevent seal jamming or losing engagement during cold-build, start-up, steady state, shutdown, and trip-shutdown
- Ensuring that the differential pressure across the seal will not cause permanent yield deformation or excessive creep over the life span
- Ensuring that there is adequate pressure load to seat and stabilize the seal
- Ensuring that leakage air flow will be sufficient to keep seal temperatures at acceptable levels
- Ensuring that cloth seal wear life will be sufficient under the combined vibration and high differential pressure load

- Identifying leakage performance improvement and integrating in to engine flow circuit

The analyses and experimentation required to fulfill these tasks include geometric analyses for engagement and jamming, finite element structural and stress analyses, rubbing wear tests, wear analyses, thermal flow analyses, subscale leakage performance tests, and analyses of leakage performance data. Fig. B7 presents an outline of cloth seal design process.

Apart from a good choice of materials and a proper weave, cloth seal design parameters include the number and thickness of the shims for structural sealing loads, the thickness of the cloth layer for sufficient wear performance, and the minimum cloth fiber diameter for adequate wear and oxidation resistance. Typically shim thickness can be determined through simple analytical calculation. In critical cases, shim stress is calculated with detailed finite-element models incorporating all the pressure, preload, and frictional contact loading. Shim stresses should be checked for worst gap and offset conditions under maximum operating pressure loads. As shown in fig. B8, for a combustor cloth seal maximum stress occurs near the pinch point.[198] Shim stresses can be determined under steady-state and transient conditions. Changes in stress due to combustion dynamics can also be determined. Analyses can be performed either by beam theory or using a finite element analysis (FEA). In both cases, shims are treated as a sequence of individual laminates made of homogenous material. Ignoring the limited additional support by the woven cloth layer will make stress analyses simpler and more conservative.

As a cloth assembly is preloaded, vibratory motion of the TD aft frame induces alternating stresses in the shims. The mean and alternating stress components can be determined using FEA. Then HCF analysis can be performed using standard formulations:[200]

$$\left(\frac{S_a}{S_e}\right)^m + \left(\frac{kS_m}{S_{ut}}\right)^p = 1$$

where S_a is the alternating stress, S_m is the mean stress, and S_{ut} is the ultimate tensile stress. Material endurance limit, S_e, needs to be modified using correction factors for surface, size, load, temperature, reliability and other like factors. Values for factor k, and exponents m and p are determined based on the fatigue-life criteria used in the analysis. If a modified Goodman fatigue-life model is selected as the failure theory, $k=m=p=1$ (see reference 200 for other failure criteria), and the above relation becomes

$$\frac{S_a}{S_e} + \frac{S_m}{S_{ut}} = \frac{1}{n}$$

where n is added in the relation to define a factor of safety against limited cycle life. For a typical Inconel X-750 shim, ultimate strength, S_{ut}, is around 1 GPa (~145 ksi) while endurance limit, S_e, is 130 MPa (~19 ksi). When corrected for surface, size, load, and temperature factors, the modified endurance stress easily reduces to S_{em} = factors × S_e = 85 MPa (12.3 ksi). Mean and alternating stress levels, S_m and S_a, are determined by the seal design and load levels. Typically, S_m is dictated by the differential pressure load and the deformation due to preload or thermal distortions (if any), and S_a is dictated by the level of pressure oscillations and relative vibration levels of the mating parts. These pressure oscillations can be due to combustion dynamics as in combustor seals, or due to pressure variations caused by blade passing at shrouds. If the von Mises stress levels under combined pressure and deflection loads are calculated to vary between σ_{max} = 350 MPa and σ_{min} = 300 MPa in a given seal application, then mean and alternating stress components would be S_m = 325 MPa and S_a = 25 MPa, respectively. Substituting in the modified Goodman failure model above, one can find that the factor of safety against infinite fatigue life is n = 1.62. If n is less than unity, this would mean that seal would have limited life for this application. Commonly, limited life means below 10^6 cycles. Therefore, n should be kept high as possible by reducing mean and alternating stress levels. Using multiple (thin) shims is a good means to reduce stress levels without adding much stiffness. Rather than deform and comply for uniform slot contact, stiff seals rotate and tip-toe, causing concentrated contact loads, which result in accelerated wear rates.

To ensure durability, a detailed wear analysis is needed. Wear of cloth mesh is much more complicated than contact of two solids. Wear rate of a woven cloth structure depends on weave, orientation, and vary through the thickness. Most wear models in the literature are based on the basic relationship developed by Archard and Hirst.[201]

$$W = K\left(\frac{FV}{3H}\right)t$$

where W is wear, K is the wear coefficient, F is the load, V is the rub-interface velocity, H is the material hardness and t is the time. Dividing both sides by contact area, the wear relation can be represented in terms of PV (pressure × rub-interface velocity), which represents contact severity conditions. Typically, *PV* charts are available for various materials to determine wear rates. However, determining the actual contact area for a cloth mesh involves some detailed geometric analysis.

As illustrated in fig. B9, when a cloth seal is pressed against the slot surface, the actual contact area of the mesh is much smaller than the nominal slot engagement area.[199] Both flat wire sections and ellipsoidal corner areas should be included in the contact area calculations (see fig. B10). As wear progresses, the actual mesh contact area changes with wear depth. For a Dutch twill weave actual contact area can be calculated as

$$A_m = \left[L2\sqrt{h(2R-h)} + \pi\left(\frac{h}{\tan\alpha}\right)c \right]\left[n_{shute}\left(\frac{n_{warp}}{4}\right)A_{nom} \right]$$

where L is the length of the flat wire sections, h is the wear depth, R is the wire radius, η_{warp} is the number of warp wires per length, η_{shute} is the number of shute wires per length, and A_{nom} is the nominal contact area.[199] Variables a, c, and α are defined in fig. B10. α can be calculated from the geometry as

$$a = \frac{h}{\tan\alpha}$$

For a proper wear analysis, the "volume lost" needs to be determined. For a given wear depth, the volume lost from the mesh can be calculated as[199]

$$V_{\text{mesh}} = \left(V_{\text{flat wire}} + V_{\text{ellcor}}\right)\left[n_{\text{shute}}\left(\frac{n_{\text{warp}}}{4}\right)A_{\text{nom}}\right]$$

where

$$V_{\text{flat wire}} = L\left[R^2 \cos^{-1}\left(\frac{R-h}{h}\right) - (R-h)\sqrt{2h(R-h)}\right]$$

$$V_{\text{ellcor}} = \left(\frac{2}{3}\right)\pi ach$$

As most of the parameters affecting wear are not constant, detailed transfer functions are needed to evaluate the effects of parameter variations on seal wear life. Using a statistical approach for the design, detailed Monte Carlo simulations can be conducted. Figure B11 shows a sample analysis for estimated seal wear after 12 000 hrs.[198] This analysis takes into account for the variation of key design and operating parameters, such as slot engagement, cantilever length, preload, operating pressure, wear coefficient, rubbing velocity, etc. The parameters are assumed to vary normally within the allowed tolerance and operating condition limits. A statistical design approach allows for a better understanding of field performance variations. A statistical study also provides valuable sensitivity data.

In most cases, leakage flow also provides cooling air to some critical parts and the seal itself. When achieving a tighter leakage performance, one should also consider the temperature increase in the seal and the surrounding slot surfaces. For critical regions, detailed thermal and flow analyses of the sealing system may be necessary. In such cases, actual leakage flow rates from the seal tests are used for better accuracy. CFD methods are used to analyze whether the temperatures exceed the temperature limitations of the cloth or shims. Figure B12 shows a sample two-dimensional thermal-flow model with symmetrical boundary conditions.[202] The conduction rate through the cloth layer should be reduced because of porosity and attention given to leakage convection. Conduction rate defines change in temperature across a unit thickness of the material in question. In porous materials with increasing porosity, solid material percentage, which defines the available pathways for heat to be conducted, decreases, and thus the heat conduction rate also decreases. Therefore the conduction coefficient for the cloth fiber material cannot be used for the cloth itself. When conducting a thermal analysis for cloth, one should decrease the conduction coefficient of fiber material with the cloth porosity to define conduction coefficient of cloth layer in the analysis. In the overall heat transfer process conduction decreases, yet convection of coolant alters the overall heat into and through the mesh. The mass flow rate should be prescribed across the cloth seal based on the data obtained from the leakage performance tests. In general, sealing locations with the extreme temperature conditions are selected for the thermal flow analyses. Typically, the highest temperature cases occur with the highest inlet flow temperature and the minimum differential pressure to drive the leakage flow. Analyses indicate that the leakage flow through the cloth cools and protects the cloth layer from high slot temperatures. Leakage flow also provides a cool buffer zone against hot streaks. The cloth layer is also effective in protecting the thin structural shim inside.

5. Summary

Cloth seals offer performance improvement through parasitic leakage reduction when applied near the hot gas path in a gas turbine engine. These locations include nozzle, shroud, and diaphragm intersegment locations; nozzle and shroud interstage locations; and transition piece and first-stage nozzle junctions, see figs. 3 and B1. The flexibility introduced by cloth seals ensures a uniform slot contact over a range of relative excursions and provides reduced leakage rates. Leakage reductions up to 30% have been achieved in combustors and 70% in nozzle segments. The flow savings have been verified through field tests with General Electric Frame 7E first-stage shroud applications. The flow savings achieved in nozzle-shroud cloth seal applications translate to performance gains of up to 0.50% output increase and 0.25% heat rate reduction in industrial gas turbines. In addition to a leakage reduction, introducing flexible combustor cloth seals have demonstrated a potential service life extension of 50% or more. Combustion laboratory tests indicate a 30-35% reduction in leakage. Currently cloth seals are standard for all new Frame 6F and 7F gas turbines. They are also offered as part of an extended life kit for the older E and F class units of Frames 3 to 9.[10]

Appendix C : Oil Brush Seals

Brush seals, after proven performance in secondary flow and hot gas path sealing applications, are being considered for oil sealing and cryogenic applications in turbomachinery, fig. 66.[153, 203] Brush seals perform very well under rotor transients owing to the inherent compliance of bristles. They have been used for oil and bearing sumps in gas turbines and aircraft engines and as buffer seals in hydrogen cooled industrial generator applications, and possible vehicle applications.[17] Tighter clearances are required at these locations to avoid oil contamination of the downstream engine components, or to minimize oil consumption levels. In generator applications, the sealing challenge is accentuated by the presence of explosive cooling gas. Typically sump applications require oil mist sealing. Successful brush seal applications at front bearing applications prevent oil mist ingestion into the compressor.[154,155] Oil mist ingestion in gas turbines and aircraft engines cause compressor blade fouling, substantial compressor efficiency loss, and potential contamination of customer air supply. In other applications where a liquid medium needs to be sealed, such as hydrogen generator buffer oil seals or liquid hydrogen-liquid oxygen seals in rocket turbo pumps,[203,204] the problem gets more complicated as hydrodynamic lift prevents bristle rotor contact, and shear heating occurs.

Because brush seals are primarily contact seals, oil temperature rise and coking become major concerns in addition to the leakage rate. Although individual bristles form very small bearing surfaces, hydrodynamic forces generated by viscous sealing medium combined with high surface speeds easily lift the compliant bristle pack off of the shaft surface.[90] The amount of lift affects the seal operating clearance, which determines oil temperature rise and leakage rate. Balancing these conflicting performance criteria requires a good understanding of bristle hydrodynamic lift. High-speed oil-brush seal test data presented by Aksit et al[153] illustrate the presence of the lift (Figure C1). The data indicate a rapid bristle lift off with rotor speed, then a stabilization of the lift clearance due to a drop in oil viscosity caused by shear heating.

1. Bristle Lift-Off

When bristle-rotor interaction is considered, the inclined approach at the tip of individual bristles create small hydrodynamic bearing surfaces at brush seal bristle tips as illustrated in Figure C2.[205] In fact, an oil brush seal can be considered as a series of small thrust bearings (one at each bristle tip) with characteristic lengths of S_T as illustrated in Figure C3.[206] This characteristic length and the actual oil lift surface at a single bristle depends on the radial penetration of the oil pumped by the rotating shaft and axial pressure drop. The thin fluid film generated by hydrodynamic lift allows reduction of general Navier-Stokes equations to the well-known Reynolds equations for bearing surfaces. The ratio of the bearing width (bristle diameter in brush seal applications) to bearing length (circumferential length of the wedge) dictates how these tiny micro-bearings behave.

Depending on seal design and operating conditions bristles can be packed very tight, allowing fluid lift pressure to act only at the very tip. In this case bearing length L is characterized by the tangential bristle spacing S_T shown in fig. C3, which is an order of magnitude smaller than the bearing width B, or bristle diameter. . These scale differences allow reduction of Reynolds equations, leading to a simplified solution, which is commonly known in tribology as a "long-bearing," solution.[207,208]

In most applications bristles deflect under differential pressure load and axially bloom, losing their tight spacing near the rotor. Typical high surface speeds in turbomachinery applications pump sealing fluid strongly into the brush pack. Therefore, actual bearing length of a bristle exposed to the fluid lift pressure is much longer than the bristle spacing, S_T. Expressing the clearance h in terms of fixed height H plus flexure and bristle radii R_a, R_b, or

$$h = H + \frac{x^2}{2R_a} + \frac{y^2}{2R_b}$$

which for the case where the bristle width and characteristic spacing are of the same order becomes

$$\frac{dh}{dx} = \frac{x}{R_a}$$

and taking advantage of long bearing length, it is possible to obtain another simplified solution, which is commonly known in tribology as a "short-bearing" solution.[207,208] The short-bearing solution results in a pressure distribution as[205]

$$P - P_a = \frac{3\mu U R_b}{R_a} \frac{x}{h^2}$$

where P_a is the ambient sump pressure. Integrating over the bearing area yields the approximate hydrodynamic lift force as[205]

$$W = \frac{6\pi}{\sqrt{2}} \mu U R_b \sqrt{\frac{R_b}{H}}$$

The long-bearing pressure and lift solutions are more complex and provided by Cetinsoy et al.[206] Hydrodynamic lift force is balanced by a reaction force due to beam/bristle deflection, frictional forces, and so-called "blow-down" forces occurring due to radial pressure gradients within the bristle pack.[90] Figure C4 compares the lift force estimates by short and long-bearing theory with beam theory results. Analyses are conducted using typical turbine oil data presented in Table C1 and published experimental oil temperature rise data.[153]

Results indicate that long-bearing theory underestimates the hydrodynamic lift. On the other hand, beam theory force results are lower than short bearing theory estimates. However, when friction and blow-down forces are also considered in addition to beam theory results to represent bristle reaction forces, the short bearing solution better represents the seal behavior.

In general the lift force increases with speed, viscosity, and bristle diameter. When the lift-radial clearance increases, the hydrodynamic lift force decreases, while the bristle tip force (due to bristle bending, blow-down and frictional interactions) increases. Multiple bristle interactions and packing are not readily modeled or determined without experiments, yet contribute to brush leakage, stiffness, and durability as pressure drop increases. The seal operating clearance occurs when forces are balanced.

2. Shear Heating

The bristle lift solutions assume constant geometry and viscosity, yet experimental data (fig. C1) indicate that hydrodynamic lift stabilizes after certain shaft speed because of shear thinning of oil and geometry changes.

Oils are quite sensitive to changes in temperature. For a turbine oil, using the supplier data for coefficient $\beta=0.0294$ with $\mu_0=0.028$ Pa-s at $T_0=37.78$ °C as the reference point, the viscosity relation becomes

$$\mu = 0.028 e^{-0.0294(T-37.78)}$$

In order to calculate the average effective fluid temperature at a given rotor speed and lift clearance, a thermal energy equation needs to be solved. Based on the experimental leakage data of Aksit et al.,[153] flow rate in leakage direction (y) is taken to be around 0.4cm^3/s. With this flow rate and other properties of the fluid medium listed in Table C1, the Peclet number, which is the ratio of forced convection to heat conduction, takes a value around 15, indicating the contribution of heat conduction to energy transfer is small in comparison to convection terms,[209] For the short bearing solution, convection and viscous dissipation dominate, and the energy equation reduces to

$$\rho c_p \upsilon_y \frac{\partial T}{\partial y} = \mu \left[\left(\frac{\partial \upsilon_x}{\partial z}\right)^2 + \left(\frac{\partial \upsilon_y}{\partial z}\right)^2 \right]$$

After a long and tedious process, the temperature distribution is reached as

$$T = T_0 + \frac{1}{2\beta} \ln\left(\frac{f_1}{f_2} + f_3 \cdot f_4\right)$$

where

$$f_1 = \exp[2\beta(T_u - T_0)]$$
$$f_2 = \exp\left[\frac{(2z-H)^2}{\rho c_p (z^2 - zH)}\left(\frac{\Delta P}{w}\right)\beta y\right]$$
$$f_3 = \frac{4(uw\mu_0)^2}{[H\Delta P(2z-H)]^2}$$
$$f_4 = \exp\left[\frac{(2z-H)^2}{\rho c_p (z^2 - zH)}\left(\frac{\Delta P}{w}\right)\beta y\right] - 1$$

Using the seal data provided by Aksit et al,[153] the calculated temperature rise values, Table C2, compare well with the experimental measurements. Higher sealing pressures derive higher leakage rates and provide more cooling at the same rotor speed. Therefore, fluid temperature rise decreases with increasing pressure load (leakage).

The long-bearing solution is much more complex and found in Cetinsoy et al.[206]

3. Fiber Selection

Oil brush seals are located near bearings and sumps. Loose ceramic or metal fibers, or their wear debris can be hazardous. Therefore, nonmetallic fibers are used in oil seals. However, organic fibers are limited in temperature capability, and tend to shrink with increase in temperature. Considering the fact that oil or oil mist at bearing cavities may reach temperatures in excess of 150 °C (352 °F), bristle shrinkage may result in increased leakage. Inertness and moisture absorption rates are the other important considerations. Aramid fibers meet all these requirements. Their high strength and excellent wear resistance make them a good choice for such applications. Coupon tests indicate they have better wear rates than typical Haynes 25 fibers (Figure C5). In addition, smooth outer surface of the fibers prevents coked oil particles from attaching and sticking fibers together.

Although oil applications of brush seals are rather new, their use in gas turbine front bearing applications have proven successful. Field tests have shown leakage reduction gains and have demonstrated durability of these seals in field operation.[154,155]

References

1. Miller, M., Colehour, J., and Dunkleberg, K., "Engine Case Externals, Challenges and Opportunities," *Proceedings of the 7th International Symposium on Transport Phenomena and Dynamics of Rotating Machinery*, ISROMAC–7 edited by A. Muszynska, J.A. Cox, D.T. Nosenzo, Oct. 1998.

2. Halila, E.E., Lenahan, D.T., and Thomas, T.T., "Energy Efficient Engine High Pressure Turbine Test Hardware: Detailed Design Report," NASA CR–167955, June 1982.

3. Hendricks, R.C., Griffin, T.A., Kline, T.R., Csavina, K.R., Pancheli, A., and Sood, D., "Relative Performance Comparison Between Baseline Labyrinth and Dual Brush Compressor Discharge Seals in a T–700 Engine Test," Paper 94–GT–266, June 1994.

4. Ludwig, L.P., and Bill, R.C., "Gas Path Sealing in Turbine Engines," ASLE Trans. vol. 23, no.1, 1980, pp. 1–22.

5. Moore, A., "Gas Turbine Engine Internal Air Systems. A Review of the Requirements and the Problems," ASME Paper 75–WA/GT–1, Nov. 1975.

6. Lattime, S.B., and Steinetz, B.M., "Turbine Engine Clearance Control Systems: Current Practices and Future Directions," *Journal of Propulsion and Power*, vol. 20, no. 2, 2004, pp. 302–311.

7. Munson, J., Grant, D., and Agrawal, G. "Foil Face Seal Proof-of-Concept Demonstration Testing," AIAA Paper 2002–3791, July 2002.

8. Chupp, R.E., Ghasripoor, F., Moore, G.D., Kalv, L.S., and Johnston, J.R., "Applying Abradable Seals to Industrial Gas Turbines," AIAA Paper 2002–3795, July 2002.

9. Bill, R.C., "Wear of Seal Materials Used in Aircraft Propulsion Systems," *Wear*, 59, no. 1, 1980, pp. 165–189.

10. Aksit, M.F., Chupp, R.E., Dinc, O.S., and Demiroghu, M., "Advanced Seals for Industrial Turbine Applications: Design Approach and Static Seal Development," *Journal of Propulsion and Power*, vol. 18, no. 6, 2002, pp. 1254–1259; See also "Advanced Flexible Seals for Gas Turbine Shroud Applications," AIAA Paper 99–2827, June 1999.

11. Camatti, M., Vannini, G., Baldassarre, L., Fulton, J., and Forte, P. "Full Load Test Experience on the Instability of a High Speed Back to Back Compressor Equipped With a Honeycomb Seal," *Proceedings of the 2nd International Symposium on Stability Control of Rotating Machinery*, ISCORMA–2, edited by A. Gosiewski and A. Muszynska, MAX MEDIA, Warsaw, Aug. 2003, pp. 617–626.

12. Camatti, M., Vannini, G., Fulton, J., and Hopenwasser, F., "Instability of a High Pressure Compressor Equipped With Honeycomb Seals," *Proceedings of the 32nd Turbomachinery Symposium*, Turbomachinery Laboratories, Texas A&M University, College Station Texas 2003.

13. Hurter, J., Zierer, T, Motzkus,T.," Sealing Improvements on the GT24/GT26 Gas Turbine Fleet-Filed Feedback on Combustor Seals, " VDI-Berichte Nr. 1965, 2006, pp 221–231.

14. Shiembob, L.T., "Development of Abradable Gas Path Seals," Pratt & Whitney Aircraft, PWA–TM–5081, East Hartford Connecticut, 1974; also NASA Contract NAS3–18023, NASA CR–134689.

15. More, D.G., and Datta, A., "Ultra High Temperature Resilient Metallic Seal Development for Aero Propulsion and Gas Turbine Applications," 2003 NASA Seal/Secondary Air System Workshop, NASA/CP—2004-212963, Washington, D.C., vol. 1, pp. 359–370.

16. Layer, J., *Advanced Metallic Seal for High Temperature Applications*, NASA CP–10198, 1997, Washington, D.C., pp. 307–328.

17. Hendricks, R.C., Braun, M.J., Canacci, V.A. and Mullen, R.L., "Brush Seals in Vehicle Tribology," Proceedings of the 13th Leeds-Lyon Symp. on Tribology, Paper E-5712, Leeds, England, 1990, pp. 231–242.

18. Bagepalli, B.S., Aksit, M.F., Farrell, T.R., Gas-path Leakage Seal for a Turbine, U.S. Patent 5934687, 10 Aug., 1999.

19. Aksit, M.F., Bagepalli, B.S., Demiroglu, M., Dinc, O.S., Keelock, I., Farrell, T., "Advanced Flexible Seals for Gas Turbine Shroud Applications," AIAA Paper 99–2827, June 1999.

20. Dinc, O.S., Bagepalli, B., Aksit, M., Wolfe, C. E., and Turnquist, N., "A New Metal Cloth Stationary Seal for Gas Turbine Applications," AIAA Paper 97–2732, July 1997.

21. Tompkins, T.L., "Ceramic Oxide Fibers Building Blocks for New Applications," *Ceramic Industry Publications,* Business News Publishing, April 1995.

22. Steinetz, B.M. and Adams, M.L., "Effects of Compression, Staging, and Braid Angle on Braided Rope Seal Performance," *Journal of Propulsion and Power*, vol. 14, no. 6, 1998, pp. 934–940; also NASA TM–107504, June 1997.

23. Steinetz, B.M., "High Temperature Braided Rope Seals for Static Sealing Applications," NASA TM–107233; also *AIAA Journal of Propulsion and Power*, vol. 13, no. 5, 1997, pp. 675–682.

24. Opila, E.J., Lorincz, J.A., Demange, J.J., "Oxidation of High-Temperature Alloy Wires in Dry Oxygen and Water Vapor," *High Temperature Corrosion and Materials Chemistry V,* Electrochemical Society, Inc., Pennington, NJ, 2005, pp. 67–80.

25. Dunlap, P.H., Steintez, B.M., Curry, D.M., DeMange, J.J., Rivers, H.K., and Hsu, S.Y., "Investigation of Control Surface Seals for Re-Entry Vehicles," *Journal of Spacecraft and Rockets*, vol. 40, no. 4, 2003, pp. 570–583.

26. Steinetz, B.M., and Dunlap, P.H., "Rocket Motor Joint Construction Including Thermal Barrier," U.S. Patent no. 6,446,979 B1 (LEW 16,684–1), Sept. 2002.

27. Hendricks, R.C., Steinetz, B.M., Zaretsky, E.V., Athavale, M.M., Przekwas, A.J., Tam, L.T., Muszynska, A., and Braun, M.J. "Reviewing Turbomachine Sealing and Secondary Flows Parts A, B, C," *The 2nd International Symposium on Stability Control of Rotating Machinery, ISCORMA-2003*, MAX MEDIA, Warsaw, Aug. 2003, pp. 40–91; see also reference 115.

28. Van Zante, D.E., Stazisar, A.J., Wood, J.R., Hathaway, M.D., and Okiishi, T.H., Recommendations for Achieving Accurate Numerical Simulation of Tip Clearance Flows in Transonic Compressor Rotors, *Journal of Turbomachinery*, vol. 122, Oct. 2000, pp. 733-742.

29. Lakshminarayana, B., "*Fluid Dynamics and Heat Transfer of Turbomachinery*," John Wiley & Sons, New York, 1996.

30. Copenhaver, W.W., Mayhew, E.R., Hah, C., and Wadia, A.R., "The Effect of Tip Clearance on a Swept Transonic Compressor Rotor," *Journal of Turbomachinery*, vol. 118, no. 2, 1996, pp. 230–239.

31. Strazisar, A.J., Wood, J.R., Hathaway, A.D., and Suder, K.L., "Laser Anemometer Measurements in a Transonic Axial-Flow Fan Rotor," NASA TP–2879, Nov. 1989.

32. Wellborn, S.R., and Okiishi, T.H., "The Influence of Shrouded Stator Cavity Flows on Multistage Compressor Performance," *Journal of Turbomachinery*, vol. 121, no. 3, 1999, pp. 486–498.

33. Bohn, D., Tummers, C. and Sell, M., "Influence of the Radial Gap on the Flow Field of a 2-Stage Turbine with Shrouded Bladings, ISROMAC-11 The Eleventh International Symposium on Transport Phenomena and Dynamics of Rotating Machinery, Honolulu, HI, USA, 26Feb-2 Mar., 2006.

34. Bill, R.C., and Wisander, D.W., "Friction and Wear of Several Compressor Gas-Path Seal Materials," NASA TP–1128, Jan. 1978.

35. Bill, R.C., Wolak, J., and Wisander, D.W., "Effects of Geometric Variables on Rub Characteristics of Ti-6Al-4V," NASA TP–1835 (AVRADCOM TR 80-C-19), April 1981.

36. Stocker, H.L, Cox, D.M., and Holle, G.F., "Aerodynamic Performance of Conventional and Advanced Design Labyrinth Seals with Solid-Smooth, Abradable, and Honeycomb Lands," NASA CR–135307 (EDR9339), Nov. 1977.

37. Stocker, H.L., "Determining and Improving Labyrinth Seal Performance in Current and Advanced High Performance Gas Turbines," AGARD–CP–237 (AGARD–AR–123), Paper 13, Aug. 1978.

38. Mahler. F.H., "Advanced Seal Technology," Pratt & Whitney Aircraft Report PWA–4372 (Contract no. AD–739922), 1972.

39. Morrell, P., Betridge, D., Greaves, M., Dorfman, M., Russo, L., Britton, C., and Harrison. K., "A New Aluminum-Silicon/Boron Nitride Powder for Clearance Control Application," ITSC 98, ASM Thermal Spray Society, 1998, pp.1187–1192.

40. Chupp, R.E., Ghasripoor, F., Turnquist, N.A., Demiroglu, M., and Aksit, M.F., "Advanced Seals for Industrial Turbine Applications: Dynamic Seal Development," *Journal of Propulsion and Power*, vol. 18, no. 6, 2002, pp. 1260–1266.

41. Schmid, R.K., Ghasripoor, F., Dorfman, M., and Wei. X., "An Overview of Compressor Abradables," Proceedings of the International Thermal Spray Conference, ITSC 2000, ASM International, 2000, pp. 1087–1093.

42. Guilemany, J.M., Navarro, J., Lorenzana, C., Vizcanio, S., and Miguel, J.M., "Tribological Behaviour of Abradable Coatings Obtained by Atmospheric Plasma Spraying (APS),"

Proceedings of the International Thermal Spray Conference, ITSC 2001, ASM International, May 2001, pp. 1115–1118.

43. Ghasripoor, F., Schmid, R.K., Dorfman, M., and Russo, L., "A Review of Clearance Control Wear Mechanisms for Low Temperature Aluminum Silicon Alloys," *Proceedings of the International Thermal Spray Conference*, ITSC 1998, Nice, France, ASM International, 1998, pp. 139–144.

44. Nava, Y., Mutasim, Z., and Coe, M., "Abradable Coatings for Low-Temperature Applications," *Proceedings of the International Thermal Spray Conference*, ITSC 2001, Singapore, ASM International, May 2001, pp. 119–126, (263–268).

45. Schmid, R., "New High Temperature Abradables for Gas Turbines," Ph.D. Dissertation, Thesis 12223, Dept. of Materials, Swiss Federal Institute of Technology, Zurich, 1997.

46. Borel, M.O., Nicoll, A.R., Schlaepfer, H.W., and Schmid, R.K., "Wear Mechanisms Occurring in Abradable Seals of Gas Turbines," *Surface Coating Technology,* 1989, pp. 117–126.

47. Chappel, D., Vo, L., and Howe, H., "Gas Path Blade Tip Seals: Abradable Seal Material Testing at Utility Gas and Steam Turbine Operating Conditions," American Society of Mechanical Engineers, Paper 2001–GT–0583, June 2001.

48. Chappel, D., Howe, H., and Vo. L, "Abradable Seal Testing: Blade Temperatures During Low Speed Rub Event," AIAA Paper 2001–3479, July 2001.

49. Ghasripoor, F., Schmid, R., and Dorfman. M., "Abradables Improve Gas Turbine Efficiency," *Journal of the Inst. of Materials*, vol. 5, no. 6, June 1997.

50. Ghasripoor, F., Schmid, R., Dorfman, M., and Wei, X., *Optimizing the Performance of Plasma Control Coatings up to 850C*, Surface Modification Technologies XII, ASME International, Rosemont, IL, 1998.

51. Shell, J.D., and Farr, H.J., "Abrasive Ceramic Matrix Turbine Blade Tip and Method for Forming," U.S. Patent no. 5,952,110, Sept. 1999.

52. Ghasripoor, F., Schmid, R.K., and Dorfman, M., "Silicon Carbide Composition for Turbine Blade Tips," U.S. Patent no. 5,997,248, Dec. 1999.

53. Benoit, R., Beverly, E.M., Love, C.M., and Mack, G.J., "Abrasive Blade Tip," U.S. Patent no. 5,603,603, Feb. 1997.

54. Draskovich, B.S., Frani, N.E., Joseph, S.S., and Narasimhan, D., "Abrasive Tip/Abradable Shroud System and Method for Gas Turbine Compressor Clearance Control," U.S. Patent no. 5,704,759, Jan. 1998.

55. Johnson, G.F., and Schilke, P.W., "Alumina Coated Silicon Carbide Abrasive," U.S. Patent no. 4,249,913, Feb. 1981.

56. Pan, Y., and Baptista, J., "Chemical Stability of Silicon Carbide in Presence of Transition Metals," *Journal. American Ceramic Society*, vol. 79, no. 8, 1996, pp. 2017–2026.

57. Hutchings, I.M., "Erosion By Solid Particle Impact," *Tribology; Friction and Wear of Engineering Materials*, Edward Arnold, London, 1992, section 6.4, pp. 171–197.

58. Biesiadny, T.J., McDonald, G.E., Hendricks, R.C., Little, J.K., Robinson, R.A., Klann, G.A., and Lassow, E. "Experimental and Analytical Study of Ceramic-Coated Turbine-Tip Shroud Seals for Small Turbojet Engines," NASA TM X–86881, Jan. 1985.

59. Wei, X., Mallon, J.R., Correa, L.F., Dorfman, M., and Ghasripoor, F., "Microstructure and Property Control of CoNiCrAlY Based Abradable Coatings for Optimal Performance,"

Proceedings of the International Thermal Spray Conference, ITSC 2000, Montreal, Canada, ASM International, 2000, pp. 407–412.

60. Burcham, R.E., and Keller, R.B., Jr., "Liquid Rocket Engine Turbopump Rotating-Shaft Seals," NASA SP–8121, Feb. 1979.

61. Alford, J.S., "Labyrinth Seal Designs Have Benefited From Development and Service Experience," SAE Paper 710435, Feb. 1971.

62. Heidegger, N.J., Hall, E.J., and Delaney, R.A., "Parameterized Study of High-Speed Compressor Seal Cavity Flow," NASA CR–198504 , 1996; also AIAA Paper 96–2807, July 1996.

63. Hendricks, R.C., and Stetz T.T., "Flow Rate and Pressure Profiles for One to Four Axially Aligned Orifice Inlets," NASA TP–2460, May 1985.

64. Egli, A., "Leakage of Steam Through Labyrinth Seals," *Transactions ASME*, vol. 57, no. 3, 1935, pp. 115–122.

65. Athavale, M.M., Steinetz, B.M., and Hendricks, R.C., "Gas Turbine Primary-Secondary Flow Path Interaction: Transient, Coupled Simulation and Comparison With Experiments," AIAA Paper 2001–3627, July 2001.

66. Thomas, R.J., "Unstable Oscillations of Turbine Rotors Due To Steam Leakage in the Clearances of the Sealing Glands and the Buckets," *Bulletin Scientifique*, vol. 71, 1958; also NASA CP–2133, 1958, pp. 1039–1063.

67. Alford, J.S., "Protection of Labyrinth Seals From Flexural Vibration," ASME Paper 63–AHGT–9; also *Journal of Engineering for Power*, vol. 86, Series A, Apr. 1964, pp. 141–148.

68. Abbott, D.R., "Advances in Labyrinth Seal Aeroelastic Instability Prediction and Prevention," *Journal of Engineering for Power*, vol. 103, April 1981, pp. 308, 312.

69. Benckert, H., and Wachter, J., "Rotordynamic Instability Problems in High-Performance Turbomachinery," NASA CP–2133, Jan. 1980, pp. 189–212.

70. Childs, D.W., Baskharone, E., and Ramsey, C., "Test Results for Rotordynamic Coefficients of the SSME HPOTP Turbine Interstage Seal With Two Swirl Brakes," NASA CP–3122, Oct. 1991, pp. 165–178.

71. Muszynska, A., "The Fluid Force Model in Rotating Machine Clearances Identified by Modal Testing and Model Applications: An Adequate Interpretation of the Fluid-Induced Instabilities, Invited Lecture," *Proceedings of the 1st International Symposium on Stability Control of Rotating Machinery (ISCORMA–1)*, edited by D. Bently, A. Muszynksa, and J.A. Cox, J.A., Bently Pressurized Bearing Corp., Mendin, NV, Aug. 2001.

72. Kanki, H., Shibabe, S., and Goshima, N., "Destabilizing Force of Labyrinth Seal Under Partial Admission Condition," *Proceedings of the 2nd International Symposium on Stability Control of Rotating Machinery (ISCORMA–2)*, edited by A. Gosiewski and A. Muszynska, MAX MEDIA, Warsaw, Poland, Aug. 2003, pp. 278–288.

73. Trutnovsky, K., "Contactless Seals, Foundations and Applications of Flows Through Slots and Labyrinths," NASA TT F 17, 352, April 1977; also Beruhrungsfreie Diechtungen, Grundlagen und Anwendungen der Stromung durch Spalte und Labyrinthe, VDI-Verlag GmbH, Dusseldorf, 1964, pp. 1–300.

74. Tseng, T., McNickel, A., Steinetz, B., and Turnquest, N., "Aspirating Seal GE90 Test," *2001 NASA Seal/Secondary Air System Workshop*, NASA/CP—2002-211911/VOL1, 2002, pp. 79–93.

75. Ferguson, J.G., "Brushes as High Performance Gas Turbine Seals," American Society of Mechanical Engineers, International Inst., Paper 88–GT–182, Amsterdam, June 1988.

76. Flower, R., "Brush Seal Development Systems," AIAA Paper 90–21443, July 1990.

77. Steinetz, B.M., Hendricks, R.C., Munson, J., "Advanced Seal Technology Role in Meeting Next Generation Turbine Engine Goals," Propulsion and Power Systems First Meeting on Design Principles and Methods for Aircraft Gas Turbine Engines, sponsored by the NATO Research and Technology Agency, Toulouse, France, May 11–15, 1998, AVT-PPS-Paper no. 11, NASA/TM-1998-206961, E-11109, May 1998.

78. Hendricks, R.C., Chupp, R.E., Lattime, S.B., and Steinetz, B.M., "Turbomachine Interface Sealing," *International Conference on Materials Coatings Thin Films, ICMCTF 2005*, American Vacuum Society, Clearance Control Session A4–1, Paper 608, San Diego, CA, May 2005.

79. Childs, D.W., Vance, J.M., and Hendricks, R.C., (eds.), "Rotordynamic Instability Problems in High-Performance Turbomachinery," *NASA Conference Publications,* NASA CP–2133, 1980; NASA CP–2250, 1982; NASA CP–2338, 1984; NASA CP–2409, 1985; NASA CP–2443, 1986; NASA CP–3026, 1988; NASA CP–3122, 1990; NASA CP–3239, 1993; NASA CP–3344, 1997; and "Instability in Rotating Machinery," NASA CP–2409, 1985.

80. Hendricks, R.C., Liang, A.D., and Steinetz, B.M., (eds.), *Seals Code Development and Seal and Secondary Air Systems Workshops Conference Publications,* NASA CP–10124, 1992; CP–10136, 1993; CP–10181, 1995; CP–10198, 1996; CP–208916, 1998; CP–210472, 2000; CP–211208, 2001; CP–211911, 2002; CP–212458, 2003; and CP–211963, 2004.

81. Chupp, R.E., and Holle, G.F., "Generalizing Circular Brush Seal Leakage Through a Randomly Distributed Bristle Bed," *ASME Journal of Turbomachinery*, vol. 118, Jan. 1996, pp. 153–161.

82. Hendricks, R.C., Liang, A.D., Childs, D.W., and Proctor, M.P., Development of Advanced Seals for Space Propulsion Turbomachinery, SAE TP Series 921028, Society of Automotive Engineers, April 1992; also NASA TM–105659, 1992.

83. Dinc, S., Demiroglu, M., Turnquist, N., Toetze, G., Maupin, J., Hopkins, J., Wolfe, C., and Florin, M., "Fundamental Design Issues of Brush Seals for Industrial Applications," *Journal of Turbomachinery*, vol. 124, April 2002, pp. 293–300.

84. Holle, G.F., and Krishnan, M.R., "Gas Turbine Engine Brush Seal Applications," AIAA Paper–90–2142, July 1990.

85. Bhate, N., Thermos, A.C., Aksit, M.F., Demiroglu, M., and Kizil, H., "Non-Metallic Brush Seals for Gas Turbine Bearings," ASME Paper GT–2004–54296, June 2004.

86. Aksit, M.F., Dogu, Y., and Gursoy, M., "Hydrodynamic Lift of Brush Seals in Oil Sealing Applications," AIAA–2004–3721, July 2004.

87. Short, J.F., Basu, P., Datta, A., Loewenthal, R.G. and Prior, R.J. "Advanced Brush Seal Development," AIAA Paper 96–2907, July 1996.

88. Chen, L.H., Wood, P.E., Jones, T.V., and Chew, J.W., "Detailed Experimental Studies of Flow in Large Scale Brush Seal Model and a Comparison With CFD Predictions," *Journal of Engineering for Gas Turbines and Power* vol. 122, no. 4, 1999, pp. 672–679; also American Society of Mechanical Engineers, Paper 99–GT–218, June, 1999.

89. Carslaw, H.S., and Jaeger, J.C., *Conduction of Heat in Solids*, Oxford Press, Oxford, England, U.K., 2nd Ed. 1959, pp. 269–270.

90. Hendricks, R.C., Schlumberger, J., Braun, M.J., Choy, F.S., and Mullen, R.L.., "A Bulk Flow Model of a Brush Seal System," American Society of Mechanical Engineers, Paper 91–GT–325, June 1991.

91. Dogu, Y., Aksit, M.F., "Brush Seal Temperature Distribution Analysis, ASME-TGTI Paper GT2005–69120 American Society of Mechanical Engineers, IGTI, Reno-Tahoe, NV, June 2005.

92. Soditus, S.M., "Commercial Aircraft Maintenance Experience Relating to Current Sealing Technology," AIAA Paper 98–3284, July 1998.

93. Proctor, M.P., and Delgado, I.R., "Leakage and Power Loss Test Results for Competing Turbine Engine Seals," American Society of Mechanical Engineers, Paper GT–2004–53935, June 2004.

94. Basu, P., Datta., P., Johnson, A., Loewenthal, R., Short, J., "Hysteresis and Bristle Stiffening Effects of Conventional Brush Seals," *Journal of Propulsion for Power*, vol. 110, no. 4, 1994, pp. 569–575.

95. Holle, G.F., Chupp, R.E., and Dowler, C.A., "Brush Seal Leakage Correlations Based on Effective Thickness," *Fourth International Symposium on Transport Phenomena and Dynamics of Rotating Machinery*, Preprint, vol. A, Begell House, New York, 1992, pp. 296–304.

96. Hendricks, R.C., Carlile, J.A., Yoder, D., and Braun, M.J., "Investigation of Flows in Bristle and Fiberglass Brush Seal Configurations," *Fourth International Symposium on Transport Phenomena and Dynamics of Rotating Machinery*, ISROMAC-4, Preprint, vol. A, Begell House, New York , 1992, pp. 315–325.

97. Braun, M.J., and Kudriavtsev, V.V., "A Numerical Simulation of a Brush Seal Section and Some Experimental Results," *Transaction of the ASME*, vol. 117, Jan. 1995, pp. 190–202.

98. Turner, M.T., Chew, J.W, and Long, C.A., "Experimental Investigation and Mathematical Modeling of Clearance Brush Seals," ASME Paper 97–GT–282, June 1997.

99. Chen, L.H., Wood, P.E., Jones, T.V., and Chew, J.W., "An Iterative CFD and Mechanical Brush Seal Model and Comparisons with Experimental Results," ASME Paper 98–GT–372, June 1998; see also Journal of Engineering for Gas Turbines and Power, vol. 121, no. 4, 1999, pp. 656–661.

100. Aksit, M.F, "Analysis of Brush Seal Bristle Stresses With Pressure Friction Coupling," ASME Paper GT–2003–38718, June 2003.

101. Mahler, F., and Boyes, E., "The Application of Brush Seals in Large Commercial Jet Engines," AIAA Paper 95–2617, July 1995.

102. Chupp, R.E. Johnson, R.P., and Loewenthal, R.G., "Brush Seal Development for Large Industrial Gas Turbines," AIAA Paper 95–3146, July 1995.

103. Chupp, R.E., Prior, R.J., and Loewenthal, R.G., "Update on Brush Seal Development for Large Industrial Gas Turbines," AIAA Paper 96–3306, July 1996.

104. Bancalari, E., Diakunchak, I.S., and McQuiggan, G., "A Review of W501G Engine Design, Development and Field Operating Experience," American Society of Mechanical Engineers, Paper GT–2003–38843, June 2003.

105. Diakunchak, I.S., Gaul, G.R., McQuiggan, G., and Southall, L.R., "Siemens Westinghouse Advanced Turbine Systems Program Final Summary," American Society of Mechanical Engineers, Paper GT–2002–30654, June 2002.

106. Ingistov, S., "Compressor Discharge Brush Seal for Gas Turbine Model 7EA," *ASME Journal of Turbomachinery,* vol. 124, no. 2, Apr. 2002, pp. 301–305.

107. Hendricks, R.C., *Environmental and Customer Driven Seal Requirements,* NASA CP–10136, 1994, pp. 67–78.

108. Schweiger, F.A., The Performance of Jet Engine Contact Seals, Lubrication Engineering, vol. 19, no. 6, 1963, pp. 232–238.

109. Brown, P.F., *Status of Understanding for Seal Materials Tribology in the 80's*, NASA CP–23000-Vol-2, 1984, pp. 811–829.

110. Lebeck, A.O., *Principles and Design of Mechanical Face Seals*, John Wiley & Sons, New York, 1991.

111. Steinetz B.M., and Hendricks, R.C., "Aircraft Engine Seals," *Tribology for Aerospace Applications*, edited by E.V. Zaretsky, STLE SP–37, 1997, chapter 9.

112. Ludwig, L.P., "Self-Acting Shaft Seals," AGARD–CP–237, Paper 16, Aug., 1978; also AGARD–AR–123 and NASA TM–73890 revised, Jan. 1978.

113. Dini, D., "Self Active Pad Seal Application for High Pressure Engines," AGARD–CP–237, Paper 17, Aug. 1978; also AGARD–AR–123, July 1978.

114. Whitlock, D.C., "Oil Sealing of Aero Engine Bearing Compartments," AGARD–CP–237, Paper 7, Aug. 1978, see also AGARD–AR–123, July 1978.

115. Boyd, G.L., Fuller, F., and Moy, J., "Hybrid-Ceramic Circumferential Carbon Ring Seal," SAE Transactions, vol. 111, pt. 1, 2002, p. 522.

116. Smith, C.R., "American Airlines Operational and Maintenance Experience With Aerodynamic Seals and Oil Seals in Turbofan Engines," AGARD–CP–237, Paper 5, Aug. 1978; also see AGARD–AR–123.

117. Hendricks, R.C., Steinetz, B.M., Athavale, M.M., Przekwas, A.J., Braun, M.J., Dozozo, M.I., Choy, F.K., Kudriavtsev, V.V., Mullen, R.L., and von Pragenau, G.L., "Interactive Developments of Seals, Bearings, and Secondary Flow Systems With the Power Stream," *International Journal of Rotating Machinery*, vol. 1, no. 3–4, 1995, pp. 153–185.

118. Hendricks, R.C., et al. "Turbomachine Sealing and Secondary Flows," NASA/TM—2004-211911/PART1, PART2, and PART3; PART 1: Review of Sealing Performance, Customer, Engine Designer, and Research Issues, Hendricks, R.C.; Steinetz, B.M.; Braun, M.J.; PART 2: Review of Rotordynamics Issues in Inherently Unsteady Flow Systems With Small Clearances; Hendricks, R.C.; Tam, L.T.; Muszynska, A.; PART 3: Review of Power-Stream Support, Unsteady Flow Systems, Seal and Disk Cavity Flows, Engine Externals, and Life and Reliability Issues, Hendricks, R.C.; Steinetz, B.M.; Zaretsky, E.V.; Athavale, M.M.; Przekwas, A.J. (see also reference 25).

119. Allcock, D.C.J., Ivey, P.C., and Turner, J.R., "Abradable Stator Gas Turbine Labyrinth Seals: Part 2 Numerical Modelling of Differing Seal Geometries and the Construction of a Second Generation Design Tool," AIAA 2002–3937, July 2002.

120. Alford, J.S., "Protecting Turbomachinery From Self-Excited Rotor Whirl," *Journal of Engineering for Power,* Series A, vol. 87, Oct. 1965, pp. 333–344.

121. Alford, J.S., "Protecting Turbomachinery From Unstable and Oscillatory Flows," *Journal of Engineering for Power*, Series A, vol. 89, Oct. 1967, pp. 513, 528.

122. von Pragenau, G.L., "Damping Seals for Turbomachinery," NASA TP–1987, March 1982.

123. Vance, J., *Rotordynamics of Turbomachinery,* John Wiley & Sons, Inc., New York, 1988.

124. Childs, D.W., *Turbomachinery Rotordynamics Phenomena, Modeling, and Analysis*, John Wiley & Sons, New York, 1993.

125. Bently, D.E., Hatch, C.T., and Grissom, B., (eds.), *Fundamentals of Rotating Machinery Diagnostics*, Bently Pressurized Bearings Press, Minden, NV, 2002.

126. Temis, Y.M., and Temis, M.Y., "Influence of Elastohydrodynamic Contact Deformations in Fluid Film Bearing on High-Speed Rotor Dynamics," *Proceedings of the 2nd International Symposium on Stability Control of Rotating Machinery (ISCORMA–2),* edited by A. Gosiewski and A. Muszynska, Paper 301, 2003, pp. 150–159.

127. Chen, J.P., "Unsteady Three-Dimensional Thin-Layer Navier-Stokes Solutions for Turbomachinery in Transonic Flows", Ph.D. Dissertation, Dept. of Aerospace Engineering, Mississippi State Univ., MS, Dec. 1991.

128. Chew, J.W., "Predictions of Flow in Rotating Disk Systems Using the k-e Turbulence Model," American Society of Mechanical Engineers, Paper 88–GT–229, June 1988.

129. Chew, J.W., Green, T., and Turner, A.B., "Rim Sealing of Rotor-Stator Wheelspaces in the Presence of External Flow," *Journal of Turbomachinery*, vol. 114, April 1992, pp. 426–432; see also 439–445.

130. Graber, D.J., Daniels, W.A., and Johnson, B.V., "Disk Pumping Test," AFWAL–TR–87–2050, Sept. 1987.

131. Johnson, B.V., Daniels, W.A., Kaweki, E.J., and Martin, R.J., "Compressor Drum Aerodynamic Experiments With Coolant Injected at Selected Locations," *Journal of Turbomachinery*, vol. 113, April, 1991, pp. 272–280; see also vol. 114, April 1992, pp. 426–432.

132. Johnson, B.V., Mack, G.J., Paolillo, R.E., and Daniels, W.A., "Turbine Rim Seal Gas Path Flow Ingestion Mechanisms," AIAA Paper 94–2703, June 1994.

133. Johnson, M.C., and Medlin, E.G., "Laminated Finger Seal with Logarithmic Curvature," U.S. Patent 5,108,116, April 1992.

134. Arora, G.K., Proctor, M.P., Steinetz, B.M., and Delgado, I.R., "Pressure Balanced, Low Hysteresis, Finger Seal Test Results," NASA/TM—1999-209191, June 1999; also ARL–MR–457, AIAA–99–2686, June 1999.

135. Proctor, M.P., Kumar, A., and Delgado, I.R., "High-Speed, High-Temperature Finger Seal Test Results," *Journal of Propulsion and Power*, vol. 20, no. 2, 2004, pp. 312-318

136. Proctor, M.P., and Delgado, I.R., "Leakage and Power Loss Test Results for Competing Turbine Engine Seals," NASA/TM-2004-213049, June 2004; also American Society of Mechanical Engineers, Paper GT–2004–53935, June 2004.

137. Braun, M., Pierson, H., Deng, D., Choy, F., Proctor, M., and Steinetz, B., "Structural and Dynamic Considerations Towards the Design of a Padded Finger Seal," AIAA Paper 2003–4698, July 2003.

138. Proctor, M.P., and Steinetz, B.M., "Non-Contacting Finger Seal," U.S. Patent 6,811,154, Nov. 2004.

139. Braun, M., Pierson, H., Deng, D., Choy, F., Proctor, M., and Steinetz, B., "Non-Contacting Finger Seal Developments," NASA/CP-2005-213655, vol. 1, Nov. 2004, pp.181–208.

140. Reed, B.D. and Schneider, S.J., "Testing of Wrought Iridium/Chemical Vapor Deposition Rhenium Rocket, NASA TM-107452, 1996.

141. Dunlap, P.H. Jr. Steinetz, B.M., and DeMange, J.J., "High Temperature Propulsion System Structural Seals for Future Space Launch Vehicles, NASA/TM-2004-212907. [E14304].

142. Flower, R.F.J., Brush Seal With Asymmetrical Elements. U.S. Patent no. 5135237, Aug.1992.

143. Nakane, H., Maekawa, A., Akita, E., Akagi, K., Nakano, T., Nishimoto, S., Hashimoto, S., Shinohara, T., and Uehara, H., "The Development of High-Performance Leaf Seals." *Transaction of ASME, Journal of Engineering and Gas Turbines and Power,* vol. 126, April 2004, pp. 342–350.

144. Steinetz, B.M., and Sirocky, P.J., "High Temperature Flexible Seal." U.S. Patent no. 4,917,302, April 1990.

145. Gardner, J., "Pressure Balanced, Radially Compliant Non-Contact Shaft Riding Seal," *NASA Seal/Secondary Flows Workshop*, NASA CP–10198, vol. 1, Oct. 1997, pp. 329–348.

146. Justak, J., "Hydrodynamic Brush Seal," U.S. Patent no. 6,428,009 B2, Aug. 2001.

147. Justak, J., "Non-Contacting Seal Developments," *2004 NASA Seal/Secondary Air System Workshop*, NASA/CP-2005-213655, Sept. 2005, pp. 101–114.

148. Shapiro, W., "Film Riding Brush Seal," *2002 NASA Seal/Secondary Air System Workshop*, NASA/CP—2003-212458/vol. 1, 2003, pp. 247–265.

149. Braun, M.J., and Choy, F.K., "Hybrid Floating Brush Seal," U.S. Patent no. 5,997,004, 7 Dec. 1999.

150. Kudriavtsev, V.V., Braun, M.J., and Choy, F.K., "Floating Brush Seal: Concept Feasibility Study," NASA SBIR Contractor Report, June 1995.

151. Lattime, S.B., "A Hybrid Floating Brush Seal for Improved Sealing and Wear Performance in Gas Turbine Applications," Ph.D. Dissertation, Dept. of Mechanical Engineering, Univ. of Akron, OH, Dec. 2000.

152. Lattime, S.B., Braun, M.J., Choy, F.K., Hendricks, R.C., and Steinetz, B.M., "Rotating Brush Seal," *The International Journal of Rotating Machinery,* vol. 8, no. 2, 2002, pp. 153–160.

153. Aksit, M. F., Bhate, N., Bouchard, C., Demiroglu, M., and Dogu, Y., "Evaluation of Brush Seal Performance for Oil Sealing Applications," AIAA paper no. AIAA-2003-4695, 2003.

154. Ingistov, S., "Power Augmentation and Retrofits of Heavy Duty Industrial Turbines model 7EA," Proceedings of Power-Gen International Conference, Las Vegas, NV, 2001.

155. Bhate, N., Thermos, A.C., Aksit, M.F., Demiroglu, M., Kizil, H., "Non-Metallic Brush Seals For Gas Turbine Bearings", Proceedings of ASME Turbo Expo 2004, GT2004-54296, 2004.

156. Pope, A.N., "Gas Bearing Sealing Means," U.S. Patent no. 5,284,347, 8 Feb. 1994.

157. Hwang, M.F., Pope, A.N., and Shucktis, B., "Advanced Seals for Engine Secondary Flowpath," *Journal of Propulsion and Power*, vol. 12, no. 4, 1996, pp. 794–799.

158. Wolfe, C.E., Bagepalli, B., Turnquist, N.A., Tseng, T.W., McNickle, A.D, Hwang, M.F., and Steinetz, B.M., "Full Scale Testing and Analytical Validation of an Aspirating Face Seal," AIAA Paper 96–2802, July 1996.

159. Bagepalli, B., Imam, I., Wolfe, C.E., Tseng, T., Shapiro, W., and Steinetz, B., "Dynamic Analysis of an Aspirating Seal for Aircraft Engine Application," AIAA Paper 96–2803, July 1996.

160. Turnquist, N.A., Bagepalli, B., Reluzco, G., Wolfe, C.E., Tseng, T.W., McNickle, A.D., Dierkes, J.T., Athavale, M., and Steinetz, B.M., "Aspirating Face Seal Modeling and Full Scale Testing," AIAA Paper 97–2631, July 1997.

161. Turnquist, N.A., Tseng, T.W., McNickle, A.D., Dierkes, J.T., Athavale, M., and Steinetz, B.M., "Analysis and Full Scale Testing of an Aspirating Face Seal with Improved Flow Isolation," AIAA Paper 98–3285, July 1998.

162. Turnquist, N.A., Tseng, T.W., McNickle, A.D., and Steinetz, B.M., "Angular Misalignment Analysis and Full Scale Testing of an Aspirating Seal," AIAA–1999–2682, June 1999.

163. McNickel, A.D., and Etsion, I., Improved Main Shaft Seal Life in Gas Turbines Using Laser Surface Texturing, NASA/CP—2002-211911/VOL1, Oct. 2002, pp. 111–126.

164. von Pragenau, G.L., "Damping Seals for Turbomachinery," NASA CP–2372, April 1985, pp. 438–451.

165. von Pragenau, G.L., "From Labyrinth Seals to Damping Seals/Bearings," Proceedings *of the 4th International Symposium on Transport Phenomena and Dynamics of Rotating Machinery (ISROMAC–4),* CRC Press, Boca Raton, FL, 1993, pp. 234–242.

166. Yu, Z., and Childs, D., "A Comparison of Experimental Rotordynamic Coefficients and Leakage Characteristics for Hole Pattern Gas Damper Seals and a Honeycomb Seal," NASA CP–3344, May 1997, pp. 77–93.

167. Etsion, I., "A New Concept of Zero-Leakage Noncontacting Mechanical Face Seal," *Transactions of ASME Journal of Tribology,* vol. 106, July 1984, pp. 338–343.

168. Young, L.A., and Lebeck, A.O., "The Design and Testing of a Wavy-Tilt-Dam Mechanical Face Seal," *Lubrication Engineering*, vol. 45, no. 5, May 1989, pp. 322–329.

169. Flaherty, A., Young, L., and Key, B., "Seals Developments at Flowserve Corporation," NASA/CP—2004-212963, vol. 1, Sept. 2004, pp. 229–238.

170. Salehi, M., Heshmat, H., Walton, J.F., and Cruszen, S., "The Application of Foil Seals to a Gas Turbine Engine, AIAA Paper 99-2821, June 1999.

171. Heshmat, H. Compliant Foil Seal, U.S. Patent 6505837 B1 14 June 2003; 19 Aug. 2003.

172. Nalotov, Oleg, "Step of Pressure of the Steam and Gas Turbine with Universal Belt," U.S. Patent 6,632,069 B1, 14 Oct. 2003.

173. Lattime, S.B., and Steinetz, B.M., "Test Rig for Evaluating Active Turbine Blade Tip Clearance Control Concepts," NASA/TM—2003-212533, July 2003. *Journal of Propulsion and Power*, vol. 21, no. 3, 2005, pp. 552–563.

174. Dunkelberg, K., "Commercial Airplane Nacelle Component (Engine Externals) Certification and Typical Temperature Exposure," *Proceedings of the 7th International Symposium on Transport Phenomena and Dynamics of Rotating Machinery (ISROMAC–7)*, edited by A. Muszynska, JA. Cox, and D.T. Nosenzo, Bird Rock Publ., Minden, NV, Feb. 1998; also NASA/TM—2006-214329/VOL1 (NASA Advanced Subsonic Technology (AST) 028, Oct. 1998), pp. 363–380.

175. Stoner, B.L., "The Importance of Engine Externals' Health," *Proceedings of the 7th International Symposium on Transport Phenomena and Dynamics of Rotating Machinery (ISROMAC–7)*, ,edited by A. Muszynska, JA. Cox, and D.T. Nosenzo, Bird Rock Publ., Minden, NV, Feb. 1998, p. 572; also NASA/TM—2006-214329/VOL1 (NASA Advanced Subsonic Technology (AST) 028, Oct. 1998), pp.435–443.

176. Zaretsky, E.V., Hendricks, R.C., and Soditus, S., "Weibull-Based Design Methodology for Rotating Aircraft Engine Structures," *Proceedings of the 9th International Symposium on Transport Phenomena and Dynamics of Rotating Machinery (ISROMAC–9)*, edited by Y. Tsujimoto, Pacific Center of Thermal-Fluids Engineering, Honolulu, HI, Feb. 2002.

177. Davis, D.Y. and Stearns, E.M., "Energy Efficient Engine Flight Propulsion System Final Design and Analysis," NASA CR–168219, Aug. 1985.

178. Athavale, M.M., Przekwas, A.J., Hendricks, R.C., and Steinetz, B.M., "Development of a Coupled, Transient Simulation Methodology for Interaction Between Primary and Secondary Flowpaths in Gas Turbine Engines," AIAA Paper 97–2727, July 1997.

179. Athavale, M.M., Przekwas, A.J., Hendricks, R.C., and Steinetz, B.M., "Coupled Transient Simulations of the Interaction Between Power and Secondary Flowpaths in Gas Turbines," AIAA Paper 98–3290, July 1998.

180. Janus, J.M., "Advanced 3–D CFD Algorithm for Turbomachinery," Ph.D. Thesis, Dept. of Aerospace Engineering, Mississippi State Univ., May 1989.

181. Janus, J.M.; and Horstman, H.Z.: Unsteady Flow-Field Simulation of Ducted Prop-Fan Configurations," AIAA Paper 92–0521, Jan. 1992.

182. Campbell, D.A., "Gas Turbine Disk Sealing System Design," AGARD–CP–237 (AGARD–AR–123), Paper 18, July 1978.

183. Teramachi, K., Manabe, T., Yanagidani, N., and Fujimura, T., Effect of Geometry and Fin Overlap on Sealing Performance of Rims Seals, AIAA Paper 2002–3938, Oct. 2002.

184. Wellborn, S.R.; and Okiishi, T.H., "Effects of Shroud Stator Cavity Flows on Multistage Axial Compressor Performance," NASA CR–198536, Oct. 1996.

185. Hall, E.J., and Delaney, R.A., "Investigation of Advanced Counterrotation Blade Configuration Concepts for High Speed Turboprop Systems," *Task 5—Unsteady Counterrotation Ducted Propfan Analysis Computer Program User's Manual,* NASA CR–187125, Jan. 1993.

186. Hall, E.J., and Delaney, R.A., "Investigation of Advanced Counterrotation Blade Configuration Concepts for High Speed Turboprop Systems," *Task 5—Unsteady Counterrotation Ducted Propfan Analysis Final Report,* NASA CR–187126, Jan. 1993.

187. Hall, E.J., and Delaney, R.A., "Investigation of Advanced Counterrotation Blade Configuration Concepts for High Speed Turboprop Systems," *Task 7—ADPAC User's Manual,* NASA CR–195472 (NASA Contract NAS3–25270), April 1996.

188. Adamczyk, J.J., Celestina, M.L., and Greitzer, E.M., "The Role of Tip Clearance in High-Speed Fan Stall," *Journal of Turbomachinery*, vol. 115, no. 1, 1993, pp. 29–39.

189. Bohn, D.E., Balkowski, I., Ma, H., Tummers, C., Sell, M., "Influence of Open and Closed Shrouded Cavities on the Flowfield in a 2-Stage Turbine, with Shrouded Bladings,: ASME GT 2003–38436, ASME Turbo Expo 2003, Atlanta, GA, USA.

190. Feiereisen, J.M., Paolillo, R.E., and Wagner, J., "UTRC Turbine Rim Seal Ingestion and Platform Cooling Experiments," AIAA Paper 2000–3371, July 2000.

191. Athavale, M.M., Przekwas, A.J., and Hendricks, R.C., "A Numerical Study of the Flow-Field in Enclosed Turbine Disk-Cavities in Gas Turbine Engines," *Proceedings of the 4th International Symposium on Transport Phenomena and Dynamics of Rotating Machinery (ISROMAC–4)*, edited by W.-J. Yang and J.H. Kim, Begell House, Boca Raton, and New York, 1992, pp. 92–101. 1992.

192. Athavale, M.M., Przekwas, A.J., Hendricks, R.C., and Steinetz, B.M., "Numerical Analysis of Intra-Cavity and Power-Stream Flow Interaction in Multiple Gas-Turbine Disk-Cavities," American Society of Mechanical Engineers Paper 95–GT–325, June 1995.

193. Virr, G.P., Chew, J.W., and Coupland, J., "Application of Computational Fluid Dynamics to Turbine Disc Cavities," *Journal of Turbomachinery*, vol. 116, 1994, pp. 701–708.

194. Ho, Y.H., Athavale, M.M., Forry, J.M., Hendricks, R.C., and Steinetz, B.M., "Numerical Simulation of Secondary Flow in Gas Turbine Disc Cavities, Including Conjugate Heat Transfer," American Society of Mechanical Engineers Paper 96–GT–67, June 1996.

195. Athavale, M.M., Ho, Y.H., and Przekwas, A.J., "Analysis of Coupled Seals, Secondary and Powerstream Flow Fields in Aircraft and Aerospace Turbomachines," Final Report, NASA Contract NAS3–27392, Final Report, Dec. 1999.

196. Smout, P.D., Chew, J.W., and Childs, P.R.N., "ICAS–GT: A European Collaborative Research Programme on Internal Cooling Air Systems for Gas Turbines," Paper GT–2002–30479, June 2002.

197. Aksit, M.F., Bagepalli, B.S., Burns, J., Stevens, P., and Vehr, J., "Parasitic Corner Leakage Reduction in Gas Turbine Nozzle-Shroud Inter-Segment Locations," AIAA Paper 01–3981, July 2001.

198. Aksit, M., Bagepalli, B., and Aslam, S., "High Performance Combustor Cloth Seals," AIAA Paper 00–3510, July 2000.

199. Ongun, R., Aksit, M.F., and Goktug, G., "A Simple Model for Wear of Metal Cloth Seals," AIAA Paper 04–3892, July 2004.

200. Avallone, A.E. and Baumeister, T., Marks' Standard Handbook for Mechanical Engineers, McGraw-Hill Book Co., 9th Ed., 1987, pp. 8–52.

201. Archard, J.F. and Hirst, W., The Wear of Metals Under Unlubricated Conditions, Proc. Roy. Soc., A236, pp. 397–410, 1956.

202. Dogu, Y., Aksit, M.F., Bagepalli, B., Burns, J., Sexton, B., and Kellock, I., "Thermal and Flow Analysis of Cloth-Seal in Slot for Gas Turbine Shroud Applications," AIAA Paper 98–3174, July 1998.

203. Hendricks, R.C., Braun, M.J., Mullen, R.L.,"Brush Seals Configurations for Cryogenic and Hot Gas Applications, Advanced Earth-to-Orbit Propulsion Technology 1990, NASA CP–3092, Vol. II, 1990, pp. 78–90.

204. Proctor, M.P., Walker, J.F., Perkins, H.D., Hoopes, J.F., Williamson,G.S., "Brush Seals for Cryogenic Applications Performance, Stage Effects, and Preliminary Wear Results in LN2 and LH2," NASA TP–3536, October 1996.

205. Aksit, M.F., Dogu, Y., J.A. Tichy and Gursoy M., "Hydrodynamic Lift of Brush Seals In Oil Sealing Applications," Proceedings of 40th AIAA/ASME/SAE/ASEE Joint Propulsion Conference & Exhibit, Fort Lauderdale, Florida, AIAA Paper AIAA–2004–3721, 2004.

206. Cetinsoy, E., Aksit, M.F., and Kandemir, I., "A Study Of Brush Seal Oil Lift Through Long Bearing Analysis," Proceedings of IJTC2006 STLE/ASME International Joint Tribology Conference, San Antonio, TX, ASME Paper IJTC2006–12370, 2006.

207. Bhushan, B., Modern Tribology Handbook, CRC Press, New York, 2001.

208. Harnoy, A., Bearing design in machinery: engineering tribology and lubrication, Marcel Dekker, 2003, pp. 148.

209. Duran, E.T., Aksit, M.F., and Dogu, Y., "Effect of Shear Heat on Hydrodynamic Lift of Brush Seals in Oil Sealing," Proceedings of 42nd AIAA/ASME/SAE/ASEE Joint Propulsion Conference & Exhibit, Sacramento, California, AIAA Paper AIAA–2006–4755, 2006.

Biographies

Raymond Chupp's career spans nearly 40 years in gas turbine design and development. In his current mechanical engineering position at GE Global Research, he has led several efforts to develop abradable tip seals for GE Energy gas-turbine product line. Previous work includes developing brush and other type advanced seals to significantly reduce leakage for expendable and long-life gas turbine engines; design of internal flow systems for various advanced aircraft and industrial gas turbines; leading experimental studies of impingement heat transfer and applying the results to airfoil design; and developing a thermal remote sensing technique for semitransparent materials. Raymond Chupp received his undergraduate degree from Kettering University and his MS and Ph.D degrees from Purdue University. He has worked at four different gas turbine manufactures during his career. He has authored 34 publications in gas turbine sealing and heat transfer, and has been granted 9 patents, with additional applications being reviewed.

Robert Hendricks began his career solving combustion problems in the NACA X-15 LOX-ammonia rocket engine. His attention then turned to fluid hydrogen heat transfer data used in all LOX-Hydrogen engine designs. He also provided fundamental understanding for boiling, two-phase flows, supercritical and near-critical fluid behavior and produced the thermophysical property codes GASP and WASP. Analyzed and validated cryogenic two-phase choked flows and the extended theory of corresponding states for fluid flow. His data showed SSME failures were caused by seal instabilities leading to research in sealing and rotordynamics. This work on a variety of seals and secondary cavity flows resulted in the design codes SCISEAL and INDSEAL and conclusively demonstrated that small changes in leakage can enhance engine performance by altering flows throughout the entire engine, prompting interactive analyses of turbomachine sealing with demonstrated increases in engine performance. His research work also includes TBC's, component life prediction, trapped vortex combustors, water injection of turbomachines. He has authored some 300 publications, and has received several NASA and paper presentation awards.

Scott Lattime has over nine years of experience as a research engineer in the industrial rotating machinery and aerospace markets with a focus on advanced bearing and seal design. For the past five years, Scott Lattime has been a member of the Seal Team at the NASA Glenn Research Center. His research included the development of innovative seals, actuators, and kinematic designs to manage blade tip clearance in the high-pressure turbine for aero-based gas turbine engines. Presently, Scott Lattime is a Product Development Specialist at the Timken Research Center in Canton, Ohio where his research is focused on the development of advanced seals, bearings, and power transmissions for new products supporting the industrial rotating machinery market. Scott Lattime earned both a Master's Degree and doctorate in Mechanical Engineering from the University of Akron. He has authored over twenty technical papers including six journal publications and one book chapter.

 Bruce Steinetz serves as NASA Glenn's Seal Team Leader. He technically directs the development of advanced turbomachinery seals (shaft seals, active mechanical tip clearance control, regenerative seals, etc), hypersonic engine and re-entry vehicle seals, and vehicle docking and berthing seals for NASA's Space Exploration Initiative. Notable projects include development and test of seals for the X-38 and X-37 (test vehicles for future re-entry systems); solid rocket motor thermal barriers (Shuttle and Atlas V), and evaluation of Shuttle Discovery main-landing gear door seals for return-to-flight. Bruce Steinetz patented and led the development of the carbon thermal barrier that prevents superheated (5500 °F) rocket combustion gas within the Space Shuttle and Atlas V motors from reaching the temperature-limited elastomeric O-rings. The Glenn thermal barriers have successfully flown on three Atlas V missions helping enable safe delivery of three satellites into orbit. The Glenn thermal barriers will be flown on future Shuttle missions starting with STS-122. Bruce Steinetz received his degrees from Case Western Reserve University. His career at NASA spans 21 years. Bruce Steinetz has been granted 9 advanced seal patents.

Figure 1.—Effects of case cooling on HPT blade tip clearance during takeoff.[2]

Figure 2.—Key aero-engine sealing and thermal restraint locations.[9]

Figure 3 (a).—Advanced seals locations in a Frame 7EA gas turbine.[10]

Figure 3 (b).—Advanced cloth seals locations in a Frame 7EA gas turbine.[10]

Figure 3 (c).—Overview of sealing in large industrial gas turbines (from ref. 13).

Figure 4.—Compressor cross-sectional drawing showing detail of rotor and seals. (a) Impeller shroud labyrinth seal. (b) Honeycomb interstage seal. (c) Abradable seal. (d) Honeycomb interstage seal.[12]

Figure 5.—Engine schematic showing main-shaft seal locations.[4]

Figure 6.—Compressor sealing locations. (a) Blade tip and interstage. (b) Drum rotor.[4]

Figure 7.—General Electric's H System gas turbine, showing an 18-stage compressor and 4-stage turbine.

Figure 8.—Temperature and pressure profiles of a Rolls-Royce Trent gas turbine engine.[*]

Figure 9.—Some types of metallic seals used in turbomachinery (courtesy of Advanced Products, Parker Hannifin Corp.).

Figure 10.—Typical gas turbine seal locations (courtesy of Advanced Products, Parker Hannifin Corp.).

Figure 11.—U-Plex and E-seal geometry. (a) Springback comparison. (b) U-Plex application.[16]

Figure 12.—Structural seal assembly.[10,15] (a) Enhanced compliance. (b) Wrapped cloth. (c) Alternate crimped cloth.

Figure 13.—Cross section of PW F119 engine showing last stage turning vane with hybrid braided rope seal around perimeter.[26]

Figure 14.—Tip flow structure for an unshrouded compressor.[27]

Figure 15.—Tip flow structure for an unshrouded turbine.[27]

Figure 16.—Contours of axial velocity (m/s) on 92 percent span stream surface from LDV measurements of a transonic compressor rotor (no frame dependencies).[28]

Figure 17.—Visualization of primary and induced clearance vortices in a transonic compressor rotor. (a) Axial velocity at 6 percent of clearance gap height from shroud. (b) Projection of relative velocity vectors on Z-r cutting plane as viewed in positive u direction, colored by u-component of velocity. Suction surface at right edge of figure. Note that the velocity profile, upper left of figure, is valid only for cascades.[28]

Figure 18.—Illustration of types of materials for interface outer air sealing. (a) Abradable (sintered or sprayed porous materials). (b) Compliant (porous material). (c) Low shear strength (sprayed aluminum).[4]

Figure 19.—Inner shrouds for compressor labyrinths. (a) Striated. (b) Honeycomb. (c) Porous material (abradable or compliant).[4]

Figure 20.—Aluminum-Silicon–polyester coating wear map using a 3-mm (0.12-in.) thick titanium blade at ambient temperature.[46]

Figure 21.—Fan shroud rub strips. (a) Potted honeycomb PW4090 fan shroud. (b) PW2000/4000 fan and rub strip interface (courtesy Sherry Soditus, United Airlines Maintenance, San Francisco, CA).

Figure 22.—Mid-temperature abradable coating (CoNiCrAlY) wear map at 500 °C (930 °F) using titanium blades.[49]

Figure 23.—Erosion and abradability as a function of ultimate tensile strength[47] (courtesy Technetics Corp.).

Figure 24.—Abradability of high temperature materials by SiC tipped blades at 1025 °C (1880 °F). Right set of data (solid black bar) is for a CaF abradable; other data are for YSZ with polyamide. The x-axis lists velocity and incursion rates. The legend gives porosity levels and average pore sizes after the polyamide is burnt out.[56]

Figure 25.—Schematic of ceramic-coated shroud seal.[58]

Figure 26.—Generalized labyrinth seal configurations.[60]

Figure 27.—Drum rotor labyrinth sealing configurations. (a) Compressor discharge stepdown labyrinth seal.[61] (b) Turbine interstage stepped labyrinth seal and shrouded-rotor straight labyrinth seal.[4]

Figure 28.—Generalized schematics of labyrinth seal throttle configurations.[63]

Figure 29.—Discharge coefficient as a function of knife-edge tooth shape.[38]

Figure 30.—Inner and outer labyrinth air seals. (a) Damper ring. (b) Damper drum (sleeve).[68] (c) Effect of seal component support.[4]

Figure 31.—Typical swirl brake configurations applied at the inlet to a labyrinth seal. (a) Radial swirl brake. (b) Improved swirl brake.[70]

Figure 32.—Labyrinth circumferential flow blocks. (a) Annular seal. (b) Labyrinth seal. (c) Flow blocks.[72]

Figure 33.—Web seal with circumferential flow blocking slots. (a) Conceptual web seal sketch. (b) Photograph of web seal.[12]

Figure 34.—Aspirating seal labyrinth tooth and seal dam sharp-edge flow restrictor. (a) At shutdown phase. (b) At steady-state operation.[74]

Figure 35.—Primary labyrinth throttle confining flows to the honeycomb journal land.[12]

Figure 36.—Typical brush seal configuration and geometric features.[40,83]

Figure 37.—Brush seal design for steam turbine applications.[40,83]

Figure 38.—Bristle stress/deflection analysis.[40,83]

Figure 39.—Brush seal performance as compared to labyrinth seal. Representative brush seal leakage data compared to a typical, 15-tooth, 0.5 mm (20 mil) clearance labyrinth seal. Measured brush seal leakage characteristic with increasing and decreasing pressure drop compared to a typical, 6-tooth, 0.5 mm (20 mil) clearance labyrinth seal.[40]

Figure 40.—Measured brush seal leakage for interference and clearance conditions.[40]

Figure 41.—7EA Gas turbine high-pressure packing brush seal in good condition after 22,000 hr of operation. [40,83]

Figure 42.—Shaft riding or circumferential contact seal.[107]

Figure 43.—Positive contact face seal.[109]

Figure 44.—Pressure balancing forces in face sealing.[111]

Figure 45.—Self-acting face seal with labyrinth seal presealing.[114]

Figure 46.—Component schematic Rayleigh pad self-acting face seal.[112]

Figure 47.—Spiral groove sealing schematic.[112]

Figure 48.—Comparison of leakage characteristics for labyrinth, conventional (contact) face seal and self-acting face seals.[112]

Figure 49.—Expanding ring seal.[109]

Figure 50.—Hybrid ceramic carbon ring seal.[115]

Figure 51.—Schematic of aero-gas-turbine buffer sealing of oil cavity.[112]

Figure 52.—Schematic of buffer fluid use in system sealing.[107]

Figure 53.—Typical multistage turbine cavity section. (a) Energy Efficient Engine high-pressure turbine.[2] (b) Hypothetical turbine secondary-air cooling and sealing[119] (courtesy AIAA).

Figure 54.—Finger seal and detailed components.[135]

Figure 55.—Illustration of a non-contacting finger seal downstream padded finger.[137]

Figure 56.—Cross section of non-contacting finger seal with two rows of padded low-pressure and padless high-pressure fingers.[138]

Figure 57.—Basic elements of leaf and wafer seals. (a) Leaf seal.[142] (b) Wafer seal.[144] (c) Canted spring preloader.

Figure 58.—Leaf seal configuration parameters. (a) Front view. (b) Side view.[144]

Figure 59.—Pressure balanced compliant film riding leaf seal.[145]

Figure 60.—Leaf-seal leakage comparison with labyrinth and brush seals.[145]

Figure 61.—Illustration of hydrodynamic brush seal (pad elements attached to bristles).[145]

Figure 62.—Hydrodynamic brush seal (spring beam elements).[147]

Figure 63.—Schematic of film riding brush seal. (a) Assembly. (b) Joint. (c) Installed.[148]

Figure 64.—The Hybrid Floating Brush Seal (HFBS).[151]

Figure 65.—HFBS performance compared to a stationary brush seal and a labyrinth seal. [$\dot{m}$ is the mass flow rate of air (pps), T_{ave} is the average upstream air temperature (°R), P_u is the average upstream air pressure (psia), D is the shaft outer diameter (in.)].[151]

Figure 66.—Sample oil brush seal with nonmetallic bristles.[153]

Figure 67.—Microdimpled surface by laser texturing.[163]

Figure 68.—Wave face seal.[169]

Figure 69.—Proof-of-concept foil face seal[7] (courtesy Rolls-Royce/Allison).

Figure 70.—Foil seal (a) schematic illustrating foil and bump-foil support [170,171] (b) foil seal "Nozzle-inlet or L-shaped" interface at attached and free end.

Figure 71.—Turbine shroud ring for deposit control. (a) Deposits build up in turbine passage. (b) Shroud discharge hole locations.[172]

Figure 72.—Thermal active clearance control system. (a) Scoop design. (b) HPT impingement manifold[2] (FADEC: full authority digital engine controller).

Figure 73.—High-pressure turbine blade tip clearance over given mission profile.[6]

Figure 74.—Effect of component temperature on predicted mean time between failures for typical engine-mounted electronic device[174] (courtesy The Boeing Company).

Figure 75.—Engine operating envelope[175] (courtesy Pratt & Whitney).

Figure 76.—Flight profile[175] (courtesy Pratt & Whitney).

Figure 77.—Material life as a function of temperature relation[175] (courtesy Pratt & Whitney).

Figure A1.—Generic turbine nozzle rotor gap configuration.[182]

Figure A2.—Experimental rim seal configurations[182] (courtesy AIAA).

Figure A3.—Comparison of experimental rim seal data at Reynolds numbers (a) 2.4×10^6 (b) 1.1×10^6 (courtesy AIAA).[183]

Figure A4.—Schematic of typical high-speed axial compressor with close-up view of seal cavity region under inner-banded stator.[62]

Figure A5.—Contours of radial velocity located one computational cell above hub (rotor) surface for course mesh. (a) Unsteady and (b) mixing plane rotor/stator/rotor ADPAC solutions.[64]

Figure A6.—Shrouded turbine configurations B and C.[35] (a) Configuration B, with closed radial gap. (b) Configuration C, with open radial gap.

Figure A7.—Pressure distribution and secondary flows at shroud: configuration C.[33]

Figure A8.—Pressure distribution and secondary flows at hub: configuration.[33]

Figure A9.—Flow domain and conjugate heat transfer calculations of all inner disk cavity pairs. Shaded areas denote conjugate heat transfer. Static pressures are specified at six main flow exits.[195]

Figure A10.—Temperature field in fluid and solid parts of turbine cavities (absolute frame).[195]

Figure A11.—Details of streamlines and temperatures in stage 1-2 cavities with conjugate heat transfer (absolute frame).[195]

Figure A12.—Locations of pressure taps in United Technologies Research Corp. experimental rig. Dots denote steady-pressure, circles denote transient pressure measurements.[65]

Figure A13.—Computational grid in disk cavity of high-pressure rig.[65]

Figure A14.—Time-dependent cavity flows for 0.69 percent purge flow (absolute frame). (a) Time-transient pressures and (b) velocity vectors in cavity. $\eta = \mathrm{Re}_{feed}/\mathrm{Re}_{turbine}^{0.8} = 0.005$, t is time, n is cycle number, and T is cycle time.[65]

Figure B1.—Advanced flexible seals for gas turbine engines.

Figure B2(a).—Typical wrapped cloth seal.

Figure B2(b).—Crimped cloth seals.

Figure B3.—Typical combustor seal assembly.

Figure B4.—Combustor cloth seals.

Figure B5.—Dutch twill weave.

Figure B6.—Cloth seal weave orientation.

Figure B7.—Cloth seal design methodology.

Figure B8.—Finite element analysis of combustor cloth seal shim stresses.

Figure B9.—Nominal and mesh contact areas.

Figure B10.—Cloth wear geometry.

Figure B11.—Sample Monte Carlo analysis for combustor seal wear in TD aft frame slot, with 95% certainty from 33.33 to 48.15% allowable thickness.

Figure B12.—Two-dimensional thermal flow model of a nozzle-shroud intersegment cloth seal.

Figure C1.—Hydrodynamic lift of brush seal with speed.

Figure C2.—Considering brush seal bristle tip geometry as convergent wedge/bearing.

Figure C3.—Bristle spacing (S_T) characterizes oil lift region at bristle tips.

Figure C4.—Comparison of the lift force estimates by long and short bearing theories with beam theory results.

Figure C5.—Wear test results: Aramid fibers against Ni-Cr-Mo-V.

TABLE 1.—GASKET AND ROPE SEAL MATERIALS

Fiber material	Maximum working temperature	
	°F	°C
Graphite		
Oxidizing environment	1000	540
Reducing	5400	2980
Fiberglass (glass dependent)	1000	540
Superalloy metals (depending on alloy)	1300–1600	705–870
Oxide ceramics (Thompkins, 1955)[21]	1800[a]	980
Nextel 312 (62% Al_2O_3, 24% SiO_2, 14% B_2O_3)	2000[a]	1090
Nextel 440 (70% Al_2O_3, 28% SiO_2, 2% B_2O_3)	2100[a]	1150
Nextel 550 (73% Al_2O_3, 27% SiO_2)		

[a]Temperature at which fiber retains 50 percent (nominal) room temperature strength.

TABLE 2.—ABRADABLE MATERIAL CLASSIFICATION

Temperature	Location/Material	Process
Low amb 400 °C (750 °F)	Fan or LPC AlSi + filler	Castings for polymer-based materials
Medium amb 760 °C (1400 °F)	LPC, HPC, LPT Ni or Co base	Brazing or diffusion bonding for honeycomb and/or fiber metals
High 760 °C (1400 °F) –1150 °C (2100 °F)	HPT YSZ and cBN or SiC	Thermal spray coatings for powdered composite materials

TABLE 3.—ABRADABLE MATERIALS USED BY CHAPPEL ET AL.[47]

Fibermetal	Density (%)	Ultimate tensile strength, (psi)
1	22	1050
2	23	2150

Honeycomb	Hastelloy-X, 0.05-mm foil, 1.59-mm cell
Nickel Graphite	Sulzer Metco 307NS (spray)
CoNiCrAlY/hBN/PE[a]	Sulzer Metco 2043 (spray)

[a]Hexagonal boron nitride (hBN) acts as a release agent; polyester (PE) controls porosity

TABLE 4.—WEAR RESISTANCE PERFORMANCE
RANKINGS OF ABRADABLE MATERIALS[47]

Material	Abradability		Erosion
	High-speed	Low-speed	
1050-psi fiber metal	1	1	3
2150-psi fiber metal	1	1	1
Hastelloy-X honeycomb	2	3	2
Nickel graphite	3	1	2
CoNiCrAlY/hBN/PE	3	3	1

Where 1 = best and 3 = worst.

TABLE 5.—TYPICAL OPERATING LIMITS FOR
STATE-OF-THE-ART BRUSH SEALS

Differential pressure	up to 300 psid per stage	2.1 MPa
Surface speed	up to 1200 ft/s	400 m/s
Operating temperature	up to 1200 °F	600 °C
Size (diameter range)	up to 120-in.	3.1 m

TABLE 6.—E3 ENGINE FLIGHT PROPULSION SYSTEM LIFE BASED ON 1985 TECHNOLOGY AND EXPERIENCE[177]

	Service life (hr)	Total life with repair (hr)
Combustor	9,000	18,000
HPT rotating structure	18,000	36,000
HPT blading	9,000	18,000
Remainder of engine	--------	36,000

TABLE C1.—TYPICAL TURBINE OIL PROPERTIES AT 50 °C

Density	$\rho = 884.61$ (kg/m^3)
Specific heat	$c_p = 2030.5$ (J/kg-°C)
Dynamic viscosity	$\mu = 0.0195$ (Pa-s)
Kinematic viscosity	$\nu = 2.2 \times 10^{-5}$ (m^2/s)
Conductivity	$k = 0.142$ (W/m-°C)

TABLE C2.—TEMPERATURE RISE ALONG Y-AXIS
(FROM UPSTREAM SIDE TO DOWNSTREAM SIDE)
FOR DIFFERENT CASES

Rotor surface speed, u	Temperature rise of the fluid across the seal	
	$\Delta P = 48.3\text{kPa}$	$\Delta P = 89.6\text{kPa}$
0	0	0
6.2 m/s	5.4 °C	5.5 °C
12.5 m/s	8.8 °C	9.3 °C
20.5 m/s	19.9 °C	16.6 °C